BIOETHICS AND THE NEW EMBRYOLOGY

Bioethics
AND THE NEW EMBRYOLOGY
SPRINGBOARDS FOR DEBATE

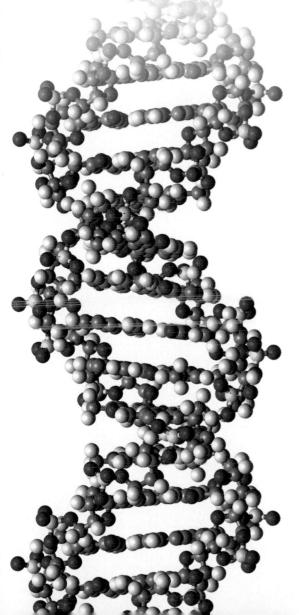

SCOTT F. GILBERT

ANNA L. TYLER

EMILY J. ZACKIN

 Sinauer Associates, Inc.

 W. H. Freeman & Company

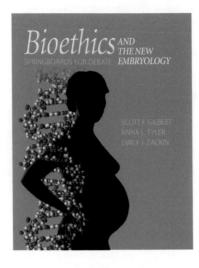

About the Cover

Cover photo © Ace Stock Limited/Alamy.
Cover design by Joanne Delphia and
Jefferson Johnson.

Bioethics and the New Embryology: Springboards for Debate

Copyright © 2005 by Sinauer Associates, Inc. All rights Reserved.

Address editorial inquiries to:
Sinauer Associates, Inc., 23 Plumtree Road, Sunderland, MA 01375 U.S.A.
publish@sinauer.com
www.sinauer.com

Address orders to: VHPS/W.H. Freeman Order Dpt.
16365 James Madison Highway, U.S. Route 15, Gordonsville, VA 22942 U.S.A.
Phone orders: 1-888-330-8477
www.whfreeman.com

Library of Congress Cataloging-in-Publication Data

Gilbert, Scott F., 1949-
Bioethics and the new embryology: springboards for debate / Scott F. Gilbert,
Anna L. Tyler, and Emily J. Zackin.
   p. cm.
ISBN-13: 978-0-7167-7345-7 (paperbound)
  1. Human reproductive technology—Moral and ethical aspects.
 2. Embryology, Human—Moral and ethical aspects. 3. Medical
ethics. 4. Stem cells—Transplantation—Moral and ethical aspects.
I. Tyler, Anna. III. Zackin, Emily.
RG133.5.G555 2005
174.2′8—dc22   2005013022

Printed in U.S.A.
5 4 3 2 1

Contents

Preface

This book was created to fill three new niches in modern college education: (1) freshman seminars that seek to unite fields of inquiry and introduce first-year college students to critical thinking about contemporary issues; (2) ethics units taught within science departments; and (3) adult education seminars that bring lifelong learners together to focus on particular topics.

One of the biggest changes in modern science has been in our ability to alter the course of human development. This has brought embryology, one of the most interesting areas of all science, into the public domain. We can plan the sex of our children in advance. We can test for the presence or absence of certain genes and abort those embryos that do not meet certain genetic specifications. We can isolate human stem cells that can generate nearly any tissue in the body, and we are able to clone human cell nuclei to produce stem cells that are genetically identical to the patients who might need them. We can already cure diseases and enhance the abilities of mice by inserting genes into their embryos. The research for applying this technology to humans has begun. The questions now may become: Even if we *can* do such things, *should* we do such things? Under what conditions should such procedures be allowed or forbidden? Do we want to support the research that might make such procedures possible?

We believe that in order to make informed judgments about such issues, one must have some background in both biology and ethics. "Facts" are important. Otherwise, what you have is a course in "Emotional Issues in Embryology." The goal of this book is to present enough science so that the reader will be able to make an informed analysis of the issues and will be able to judge whether their views are concordant with what we know about embryos and cells. Toward this end, each of the six two-chapter units pairs a chapter covering basic science with a chapter covering various views on the ethical questions raised by these scientific issues.

As the book's subtitle states, these chapters were written to be springboards for discussion. They certainly are not expected to provide definitive answers. There are some biases in this book. First, the senior author, Scott Gilbert, received his Ph.D. in biology from the Johns Hopkins University, where he pursued research on embryonic human cells with the goal of devising ways to diagnose certain genetic diseases. He favors the legality of abortion. He also has a B.A. degree in religion and an M.A. in the history of science. Scott has written that the arguments about cloning, stem cells, and other issues do not fall along a "science versus religion" divide, as is so often represented in the news media. Rather, there are scientists who vehemently oppose certain technologies and religious figures who enthusiastically welcome them. Often, what is at stake involves competing notions of human dignity, not necessarily a "scientific" or "religious" worldview.

Second, the papers herein presuppose the existence of a capitalistic health care system working within a democratic society. The book deals with issues where scientific knowledge plays a role in the health care debate. It does not deal with such extremely important issues as whether medicine should be a for-profit enterprise, who cares for children born with life-threatening abnormalities, the commercialization of an individual's tissues, the legal ability to refuse genetic testing without negative consequences, or whether malpractice suits have changed how pregnant women receive medical care. These are critical issues, but they deal not so much with science as with other social agendas.

Third, the book presumes that knowledge of the science is important for making informed decisions about public policy. The opening quotation from Thomas Huxley, a Victorian biologist who was also an educator, philosopher, and health care reformer, is very pertinent to these chapters. Most of the arguments presented in the popular media are done in the absence of the science involved. Indeed, much of the way science is reported seems calculated to elicit emotional responses. We hope to be able to look behind the rhetoric to see what the scientific data are.

Although the science presented in the book has been simplified and explained at the level of an introductory biology course, it still should convey the necessary information to encourage useful discussions. We are introducing relevant material in biology and ethics, not exploring all its nuances. There are plenty of topics for students to develop into term papers, whether these be on the physiology of stem cells or on the philosophy of essentialism.

The book follows the same sequence of topics that most developmental biology courses have used to introduce this material. Thus, the first section starts with an overview of human development, from fertilization to birth.

The ethical chapter asks the question, "When does human life begin?" The second section focuses on fertilization, and the question becomes whether modern assisted reproductive technologies should be regulated more closely. The third section continues with an aspect of fertilization—the genetic determination of sex—and then looks at whether we should be allowed to use newly available fertilization technologies to "pre-select" the sex of our children. The fourth section looks at the science of cloning, from early experiments with sea urchins and amphibians to present work on mammals, and asks whether reproductive cloning techniques should be applied to humans. The fifth section concerns the formation of tissue and organs from pluripotent embryonic stem cells; the ethical issue here is whether to use such embryonic stem cells to make new tissues to replace diseased or missing tissues. The sixth section concerns the mechanisms by which different genes are expressed in different cell types. The ethical issues associated with this involve genetic enhancement—whether we have the right (or even the obligation) to modify our progeny's genes to make them more fit in our environment. Although most of these chapters contain biomedical information, we have attempted to present it at a level for general audiences. The information here is not intended to be a substitute for professional medical or genetic counseling.

The book ends with a group of three chapters covering issues that predate modern reproductive and stem cell technology but that are certainly highlighted by our new abilities. The first is the question of "What is normal?"; the second discusses what genes can and cannot do, and how the media has often distorted our view of gene function. The last chapter concerns the use of animals in our laboratories.

Many of these subjects—sex selection, genetic enhancement, cloning, regenerating replacement tissues—were once the realm of science fiction authors. This fiction has become reality; the future is here. But the form of our future is still malleable, and the debates that will take place over the next few years may be critical in giving it

> The future is already here.
> It's just not evenly distributed yet.
>
> WILLIAM GIBSON (1999)

shape. It will be a very interesting time, and we hope this book will allow its readers to better appreciate what is going on.

Acknowledgments

This book began as part of a "History and Critiques of Biology" course that I taught in 2002. I asked my students to organize themselves into groups that would define the scientific and cultural concerns involving several areas where embryology met the larger society. These students included K. Cloonan, M. Cho, C. Crumley, S. Kiymaz, N. Schichor, J. Simonet, C. Celano

A. Tyler, R. Messing, M. Cohen, S. Sistla, J. Miyamae, A. M. Lam, K. Davenport, B. Davis, P. Riccio, H. Fleharty, C. Withers, J. Martin, and M. Hashimoto. These students ranged through all four classes, several different majors, and a wide range of cultural perspectives. Emily Zackin, a senior who was majoring in political science and English literature with a minor in biology, was the perfect teaching assistant to help these students find materials in a variety of sources.

When I read what these students put together, I believed that I might have the basis for a textbook more useful than any I had previously seen on these topics. So we put the papers on the Web. Anna Tyler, a former student in my course, worked with me to update and revise the chapters for this book.

Others have helped enormously. The material on stem cells was reformulated and updated by Mary Tyler of the University of Maine. Anna Tyler was responsible for the initial drafts of the animal use chapter, and this was further edited by me and by Donna Haraway. These chapters were then given to adult students in a lifetime learning seminar on bioethics and embryology that I coordinated in 2004, and also to a panel of reviewers that included biologists Yolanda Cruz and Judith Cebra-Thomas, and bioethicist Paul Root Wolpe. Anne Raunio, M.D. provided expert reviewing for sections concerning obstetrics and gynecology. The comments of these advisors have been extremely useful, but they should not be held responsible for the material herein. The text was given a final editing by me and by Carol Wigg of Sinauer Associates. Carol became a "fourth author," and we thank her enormously for her suggestions on organization and for making sure that the language was appropriate for the audiences we hope to reach. We also thank Andy Sinauer and Dean Scudder for their enthusiasm about publishing this book and Jefferson Johnson for his creative design.

Scott Gilbert

UNIT 1

When Does Human Life Begin?

The history of a man for the nine months preceding his birth would probably be more interesting, and contain events of far greater moment, than the three-score and ten years that follow it.

SAMUEL TAYLOR COLERIDGE (1885)

CHAPTER 1

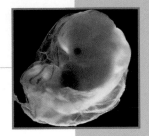

An Outline of Human Development

The eminent Victorian Samuel Taylor Coleridge got it right. The concept of an embryo is a staggering one. To become an embryo, you had to build yourself from a single cell. You had to respire before you had lungs, digest before you had a gut, build bones when you were pulpy, and form orderly arrays of neurons before you knew how to think. One of the critical differences between you and a machine is that a machine is never required to function until after it is built. Every animal has to function as it builds itself.

The only way for an animal to grow from a single cell—the fertilized egg—into a complex, many-celled adult is by first developing an embryo. Fertilization gives each animal a **genome**, a set of genes found in every cell of the animal's body. The genome contains most of the instructions needed by the embryo to grow in particular ways—it specifies the individual's **genotype**. Some of these instructions define the species: a pregnant dog will give birth to puppies, while her pregnant owner will give birth to a human baby. Other instructions encoded in the genotype produce variations within the species—a chocolate lab or an Australian shepherd; blonde, red, or black hair; a wide nose or a narrow nose. The set of observable traits produced by the genotype (and influenced by environmental factors) is called the **phenotype**. The embryo is the link between the genotype and the phenotype.

The branch of biomedical science that studies the development and transformation of embryos is called either **embryology** or by its more recent name, **developmental biology**. Whereas most of biology studies the structure and function of adult organisms, the interests of developmental biology lie in the transient stages leading up to the adult. Developmental biology studies the initiation and construction of organisms. It is a science of becoming, a science of process. To say that a mayfly lives but one day is profoundly inaccurate to a developmental biologist. A mayfly may be a winged adult for only a day, but it spends the other 364 days of its life developing into that adult beneath the waters of a pond or stream.

The Questions of Developmental Biology

The development of an animal from an egg has been a source of wonder throughout history. The simple procedure of cracking open a chick egg on each successive day of its three-week incubation provides a remarkable experience as you watch a thin band of cells give rise to an entire bird. Aristotle, the first naturalist known to have recorded embryological studies, performed this procedure and noted the formation of the bird's major organs. Anyone can wonder at this remarkable—yet commonplace—phenomenon, but the scientist seeks to discover how development actually occurs. And rather than dissipating wonder, the scientist's new understanding increases it.

> It is owing to wonder that people began to philosophize, and wonder remains the beginning of knowledge.
>
> ARISTOTLE (350 BCE)

Development is a relatively slow process of progressive change that accomplishes two major objectives: (1) it generates cellular diversity and order within each individual organism, and (2) it ensures the continuity of life from one generation to the next. Thus, embryologists deal with two fundamental questions: How does the single-celled fertilized egg give rise to the many different cells of the adult body? And how does that adult body produce yet another body? These two huge questions have been traditionally subdivided into four general questions scrutinized by developmental biologists:

1. *The question of differentiation.* A single cell, the fertilized egg, gives rise to hundreds of different cell types—muscle cells, skin cells, neurons, lens cells, lymphocytes, blood cells, fat cells, and so on (Figure 1.1). This generation of cellular diversity is called **differentiation**. Since each cell of the body (with very few exceptions) contains the same set of genes, we need to understand how this same set of genetic instructions can produce different types of cells. How can the fertilized egg generate so many different cell types? How does one part of the gut tube become the pancreas while the part next to it becomes the liver? How does one particular cell give rise to all the many different types of blood cells, while another cell gives rise to many different types of nerve cells?

2. *The question of morphogenesis.* Our differentiated cells are not randomly distributed, but are organized into intricate tissues and organs. During

(A)

(B)

(C)

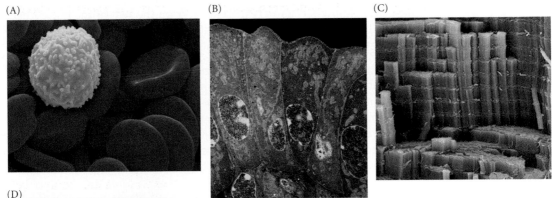

(D)

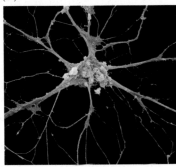

FIGURE 1.1   Some differentiated cells of the human body. (A) Red and white blood cells. Red blood cells carry oxygen to the rest of the body's cells. White blood cells, such as the single one in this picture, are part of the body's immune system and play an important role in fighting bacteria and other disease-causing organisms. (B) Epithelial cells form a sheet lining the small intestine. Epithelial cells are found throughout the body; these particular digestive-tract cells secrete enzymes that digest food. (C) Skeletal muscle cells form fibers. The individual cells are not visible here, but this micrograph gives a sense of how muscles work as the fibers slide past one another to create movement. (D) A neuron, or nerve cell. The long, thin structures radiating from the central cell transmit and receive sensory information that is processed by the brain. (Micrographs © SPL/Photo Researchers, Inc.)

development, cells divide, migrate, aggregate, and die; tissues fold and separate. The organs are formed and arranged in a particular way: our fingers are always at the tips of our hands, never in the middle; our eyes are always in our heads, not in our toes or gut. This creation of ordered form is called **morphogenesis**. How do millions of individual cells form such ordered structures?

> My dear fellow … life is infinitely stranger than anything which the mind of man could invent. We would not dare to conceive the things which are really mere commonplaces of existence.
>
> ARTHUR CONAN DOYLE, *A CASE OF IDENTITY* (1891)

3. *The question of growth.* How do our cells know when to stop dividing? If each cell in our face were to undergo just one more cell division, we would be horribly malformed. If each cell in our arms underwent just one more round of cell division, we could tie our shoelaces without bending over. Our arms are generally the same size on both sides of the body. How is cell division so tightly regulated?

4. *The question of reproduction.* In humans and other vertebrates, only certain very specialized cells—the sperm and the egg—can transmit the instructions for making an organism from one generation to the next. How are

these cells, known as **gametes**, set apart from the rest of the body's cells to form the next generation, and what special instructions do they carry that allow them to unite and form another embryo?

These questions are among the most exciting in all science because they are fundamental to how we come into being. They are also among the most exciting questions in medicine, because birth defects and cancers involve errors in the processes of growth, differentiation, and morphogenesis. And coupled with recent advances in biotechnology, knowledge of how cells grow, differentiate, and order themselves may lead to new procedures for regenerating damaged organs or enhancing existing ones.

Human Embryonic Stages

All animals come from a common ancestral origin, and the basic structure of our life cycle is not much different from those of earthworms, eagles, flies, or beagles. Some people come out of embryology classes with a new philosophy of life. Others come out of the same class merely knowing a lot more Greek than when they went in. Here is the Greek part—the vocabulary by which biologists name the stages of human development.

The generation of a new organism is initiated (but not completed) by the fusion of genetic material from two specialized cells—the sperm and the egg (Figure 1.2). This fusion, called **fertilization** (or, in humans, **conception)**, combines the parents' genes and stimulates the egg to begin development. The products of human conception are sometimes called the **conceptus**. These products include the zygote, the embryo, the embryonic sacs (amnion, chorion, yolk sac, and allantois), and the fetus.

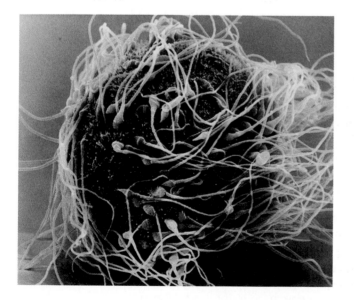

FIGURE 1.2   Human egg and sperm. This micrograph has been artificially colored so that the egg (blue) and sperm (white) can be distinguished. Notice how much larger the egg is than the sperm; the actual size of the egg is about one-tenth of a millimeter, barely visible to the human eye. Although many sperm can attach to the egg's surface, as shown here, normally only one sperm enters and fertilizes the egg. (Photograph by D. M. Phillips/Photo Researchers, Inc.)

Fertilization

We will discuss fertilization in detail in Chapter 3. Here we will simply say that the sperm brings to the egg 23 chromosomes containing the genes that the father will pass on to the new person. The sperm also provides a structure called the centriole that will be needed for cell division. The egg contributes its own genes in a set of 23 chromosomes, giving the new individual a complete complement, or **karyotype,** of 46 chromosomes (Figure 1.3). The egg—which is thousands of times bigger than the sperm—also contributes a large amount of cellular material called cytoplasm. In the egg cytoplasm are structures called mitochondria that will provide energy for embryonic development, along with other biochemical factors that instruct the early division and differentiation of the cells.

The fertilized egg is called the **zygote** (Greek for "tethered together"). The human zygote is only 100 μm (100 microns, or one-tenth of a millimeter) in diameter—barely visible to the eye.

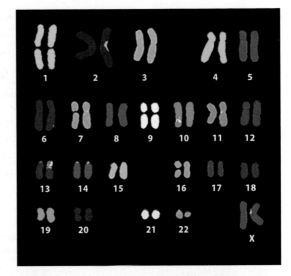

FIGURE 1.3 A human karyotype. Every human body cell has two copies each of 23 chromosomes, for a total of 46 chromosomes in its nucleus. Each of the 23 chromosomes can be stained a different color because the different genes present on them give each a distinct biochemical composition. The X and Y chromosomes determine the individual's sex, as discussed at length in Chapter 5; this person has two X chromosomes and is thus female. (Courtesy of Drs. T. Ried and E. Schröck, National Institutes of Health).

Embryogenesis

The processes that comprise development between fertilization and birth are collectively called embryogenesis. In humans, the period between fertilization and birth is called the gestation period, and it usually lasts 38 weeks (just over 9 months; an average of 266 days).

During the first 8 weeks of gestation, the conceptus is called an embryo. During this embryonic period, the organ systems begin to form and their cells differentiate. This crucial 60-day period has been divided into 23 specific stages, called "Carnegie stages" after the Carnegie Institute at Johns Hopkins University (the location of the laboratory that first described these stages in the early 1900s) (Figure 1.4). The events that take place during the Carnegie stages are intricate and sensitive; even slight biochemical or genetic aberrations can have massive effects on the emerging embryo. The vast majority of miscarriages occur during the embryonic period, often before the woman even realizes she is pregnant.

FIGURE 1.4   The Carnegie stages of the human embryonic period (the 60 days immediately following conception). (A) Stages 1–13 (approximately 30 days) cover the events of fertilization, implantation, cleavage, gastrulation, and neural tube formation. Note that the images are not to scale. **Stage 1:** The fertilized egg is 0.1–0.15 mm in diameter (about half the size of the period at the end of this sentence). **Stage 2:** Early cleavage, from the 2- to the 16-cell embryo. **Stage 3:** The blastocyst has an inner cell mass and trophoblast; is no more than 0.2 mm in diameter. This is the stage from which embryonic stem cells are obtained. At stages 4–7 (not shown), the blastocyst implants in the uterus and begins gastrulation. **Stage 8:** This model depicts embryo toward the end of gastrulation; it is now about 2 mm long. **Stage 9:** Model showing the neural folds, which indicate the beginning of neural tube formation. **Stages 10 & 11:** The neural tube fuses. The embryo at this point is no more than 3 mm long. **Stage 12:** The rostral (head) and caudal (tail) regions of the embryo are discernible.

(A)

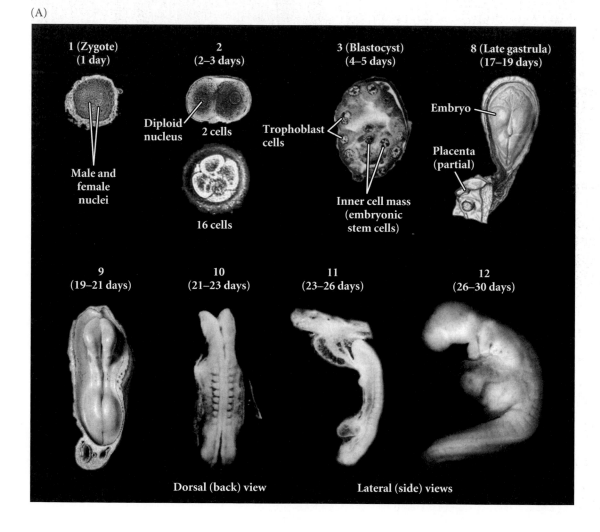

(B) Stages 14–23 (days 30–60). The body's major organs form over the remainder of the embryonic period, as the embryo grows from approximately 3 mm to about 35 mm long. After 60 days, development becomes primarily a matter of growth and maturation rather than organ formation, and the conceptus is referred to as a fetus rather than an embryo. (A, photographs and models from the Louisiana State University Heirloom Collection, courtesy of the National Library of Medicine; B, photographs from the Kyoto Collection, Kyoto University Graduate School of Medicine, courtesy of Dr. Kohei Shiota.)

(B)

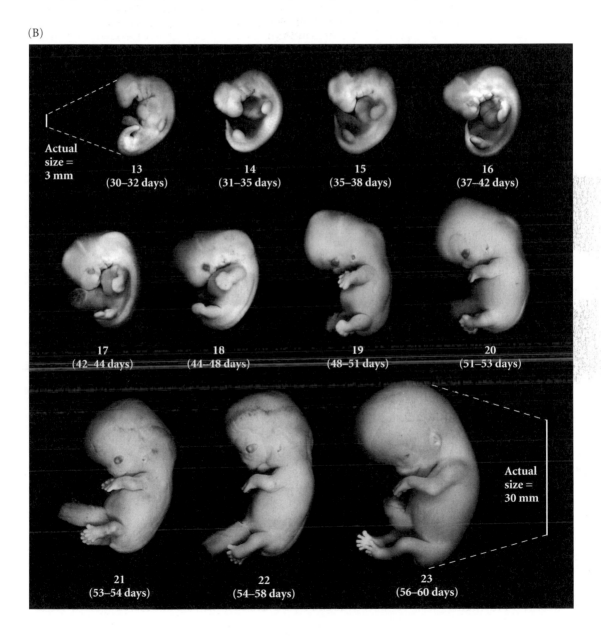

Actual size = 3 mm

13 (30–32 days)

14 (31–35 days)

15 (35–38 days)

16 (37–42 days)

17 (42–44 days)

18 (44–48 days)

19 (48–51 days)

20 (51–53 days)

Actual size = 30 mm

21 (53–54 days)

22 (54–58 days)

23 (56–60 days)

(A)

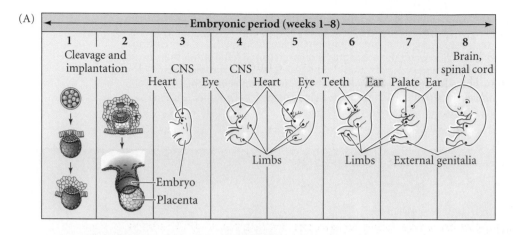

(B)

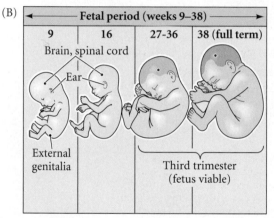

FIGURE 1.5   The periods of human development. (A) The embryonic period lasts the first 8 weeks and includes the period before the embryo implants into the uterus, as well as the stages where the major organs begin to form. (B) The fetal period, beginning on week 9 and extending to the full-term infant at approximately week 38, is mainly one of growth and maturation. The labeled dots indicate critical times for the formation of some major organ systems. During these time frames, the organs are particularly susceptible to drugs and other external agents that can cause birth defects (see pages 23–25). The central nervous system, or CNS, includes the brain and spinal cord (red dots) and develops over the full term of gestation.

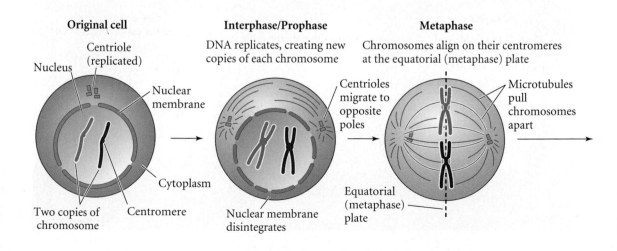

During the 30 weeks following the embryonic period, the conceptus is called a fetus. During the fetal period, the organ systems mature and the conceptus grows enormously.

Physicians often divide human gestation into three trimesters of three months each. The first trimester subsumes the embryonic period and early fetal limb and organ formation. The second and third trimesters are primarily periods of growth and maturation (Figure 1.5). The third trimester, beginning at around week 27, is historically the time when the fetus, if born prematurely, would have a reasonable chance of independent survival (i.e., survival without modern technological support).

Cleavage

The first stage of embryogenesis is **cleavage**. During cleavage, the cells divide by a process called **mitosis** (Figure 1.6). In almost all of the body's cells, there are two copies of each chromosome—one copy from the mother and one from the father (see Figure 1.3). During mitosis, the chromosomes first duplicate themselves so that the cell has four copies of each chromosome. Then the cell divides in two in such a way that each new cell contains one copy of each chromosome—that is, each of the two new cells contains the same two chromosomes as the original cell. Thus, almost every

FIGURE 1.6   During mitotic cell division, a chromosome doubles by replicating its DNA. The centriole has also duplicated, and the two centrioles now migrate to an opposite sides of the nucleus. The duplicated chromosomes become attached to the centrioles by microtubules, which pull the chromosomes apart so that one chromosome goes into each new cell. To make it easier to visualize, the process is diagrammed here for only one chromosome; however, in humans all 46 chromosomes will duplicate and then divide in the same mitotic cycle, so that the number of chromosomes in each cell remains 46.

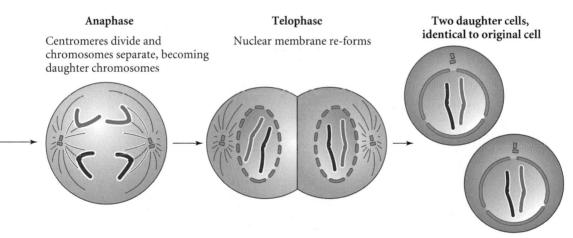

| Anaphase | Telophase | Two daughter cells, identical to original cell |
| Centromeres divide and chromosomes separate, becoming daughter chromosomes | Nuclear membrane re-forms | |

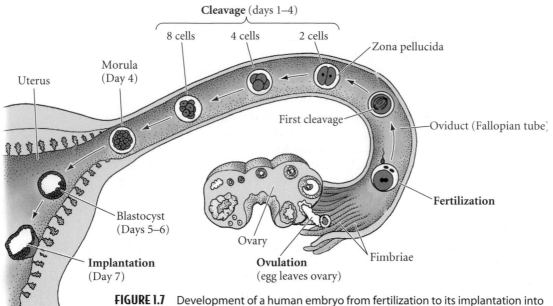

FIGURE 1.7   Development of a human embryo from fertilization to its implantation into the uterus. Compaction of the human embryo into a morula occurs on day 4, shortly after each cell has divided three times. The embryo "hatches" from the protective zona pellucida upon reaching the uterus (see Figure 1.12).

cell in your body has the same genome as every other cell.* Indeed, each body cell (**somatic cell**) is the mitotic descendant of the zygote.

Humans follow the typical mammalian pattern of cleavage, which takes much longer than those of most other animals. Given that mammalian cleavage takes place inside the mother's body, it has been very difficult to study; however, the knowledge we now possess was worth waiting for.

One of the most amazing things about mammalian cleavage is that it must be synchronized with the migration of the embryo into the uterus. The young, unfertilized mammalian egg is released from the ovary and swept into the **oviduct** (also called the **Fallopian tube**) by fingerlike projections of the oviduct called fimbriae (Figure 1.7). The fimbriae have a protein that recognizes the protein matrix surrounding the egg, and they transfer the egg into the tube adjacent to the ovary. Fertilization occurs in a region of the oviduct that is close the ovary. This is important, because if fertilization occurs too close to the uterus (womb), the zygote won't have enough

---

*Red blood cells (Figure 1.1A) and lymph cells do not carry the same genetic information found in other somatic cells. Red blood cells (also known as erythrocytes) "toss out" their gene-bearing nuclei and become full of hemoglobin, the protein that carries oxygen molecules. Lymph cells (lymphocytes) alter the genes that produce antibodies as part of our immune response.

time to make the cells that stick to the lining of the uterus and allow the embryo to implant into it.

Immediately after fertilization, the cells divide once every 12–18 hours. Each cleavage-stage cell is called a **blastomere**.

Compaction

In mammals, the first cell type differentiation occurs during the cleavage stage. This differentiation creates the cells by which the embryo will attach to the uterus, so it has to take place in the oviduct, before the embryo reaches the womb. In the oviduct, hairlike projections called cilia gently push the embryo toward the uterus; the embryonic cells continue to divide during this 5-day journey (see Figure 1.7).

Shortly after the formation of 8 cells (that is, after the third cell division, around four days after fertilization), the blastomeres remain loosely connected to one another. However, soon afterward, the blastomeres undergo a spectacular change in their behavior. They suddenly huddle together, maximizing their contact with one another and forming a compact ball of cells (Figure 1.8). This tightly packed arrangement is stabilized by tight junctions that form between the outside cells of the ball, sealing off the inside of the sphere.

The cells of the compacted embryo divide to produce a 16-cell **morula** (Latin for "mulberry," which it vaguely resembles; see Figure 1.8E). The morula consists of a small group of *internal* cells surrounded by a larger group of *external* cells. Most of the descendants of the exterior cells become

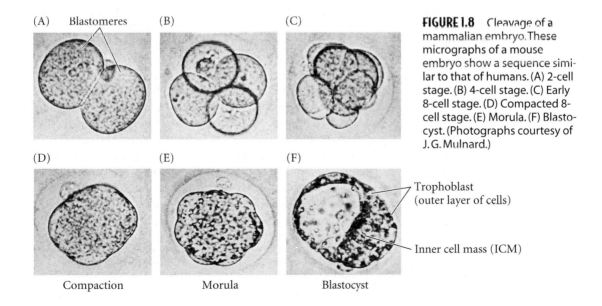

(A) Blastomeres (B) (C)

(D) (E) (F)

Compaction   Morula   Blastocyst

Trophoblast (outer layer of cells)

Inner cell mass (ICM)

FIGURE 1.8 Cleavage of a mammalian embryo. These micrographs of a mouse embryo show a sequence similar to that of humans. (A) 2-cell stage. (B) 4-cell stage. (C) Early 8-cell stage. (D) Compacted 8-cell stage. (E) Morula. (F) Blastocyst. (Photographs courtesy of J.G. Mulnard.)

FIGURE 1.9   A human conceptus at 50 days of gestation (the end of week 6). The embryo lies within the amnion, and its placental blood vessels can be seen extending into the chorionic villi (fingerlike projections of the chorion). This remarkable classic photograph shows the conceptus slightly larger than its actual size. At this point in development the embryo is in fact about 2 centimeters (less than 1 inch) long. (Photograph from the Carnegie Institution of Washington, courtesy of C. F. Reather.)

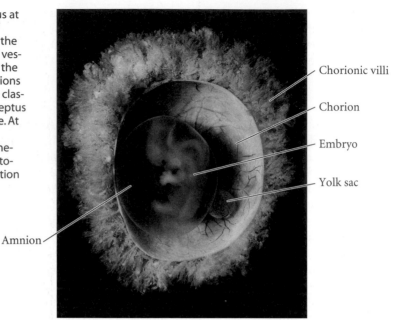

Chorionic villi

Chorion

Embryo

Yolk sac

Amnion

trophoblast cells. Trophoblast cells produce no embryonic structures. Rather, they form the tissue of the **chorion**, the embryonic portion of the **placenta** (Figure 1.9). The chorion enables the embryonic blood vessels to come into contact with the mother's blood vessels, thus allowing the embryo to receive oxygen and nourishment from its mother. The chorion also secretes hormones that cause the uterus to remain soft and to retain the fetus as it grows. In addition, the chorion produces chemicals that block the mother's immune system so that the mother will not reject the embryo.*

The embryo proper is derived from the descendants of the *inner* cells of the 16-cell stage. These cells generate the **inner cell mass (ICM)**, which will give rise to the embryo and its associated yolk sac, allantois (waste sac), and amnion (water sac) (Figure 1.9). The distinction between trophoblast and inner cell mass represents the first differentiation event in mammalian development. The trophoblast cells must be made first, since these are the cells that will adhere to the uterus. Once the trophoblast adheres to the uterus, the trophoblast and uterus interact to make the placenta. During this time, the cells of the inner cell mass secrete proteins that help the tro-

---

*Because approximately half of the embryo's genes come from its father, the embryo's tissues are different from ("foreign to") the mother's tissues. Without the chemical blockade produce by the chorion, the mother's system would reject the embryo, just as it would reject a donated organ if immunosuppressant drugs were not administered.

phoblast grow. Only later will the inner cell mass start to make the embryo. This stage, when the conceptus has an inner cell mass surrounded by trophoblast cells, is called a **blastocyst** (see Figure 1.8F).

The inner cell mass: Twins and stem cells

Before the blastocyst is formed, each of the early blastomeres (2-, 4-, or 8-cell stage) is thought to be able to form an entire embryo, including the trophoblast. This amazing ability is called **regulation**, and it stunned early embryologists. It means that if you were to take a single blastomere out of the 8-cell embryo and place that cell into a glass dish full of nutrients, that cell by itself would generate a complete blastocyst. This blastocyst could be placed into a uterus and develop into an entire organism. The isolated blastomere doesn't make one-eighth of an embryo—it can make all the cells of the body. Thus, blastomeres are said to be **totipotent** (Latin, "capable of becoming anything").

During the blastocyst stage, each blastomere of the inner cell mass has the ability to form any type of cell in the embryo proper, but the isolated ICM cells probably cannot form trophoblast cells. Thus, cells of the inner cell mass are said to be **pluripotent** (Latin, "capable of becoming many things"; see Chapter 7). In other words, each of the cells of the inner cell mass is not yet determined to become any specific type of cell, and what cell type they will become depends largely upon interactions between cells. These experiments have been done with mouse embryos, but whether this restriction in potency applies to human embryos is not yet known. However, the regulative capacity of the ICM cells is seen in human twinning.

Human twins are classified into two major groups: **monozygotic** (Greek, one-egg) or identical twins, and **dizygotic** (two-egg), or fraternal twins. Fraternal twins are the result of two separate fertilization events and have separate and distinct genotypes, whereas identical twins are formed from a single embryo whose cells somehow dissociated from one another, and thus share a common genotype (Figure 1.10). Identical twins may be produced by the separation of early blastomeres, or even by the separation of the inner cell mass into two smaller clusters within the same blastocyst.

Identical twins occur in roughly 0.25 percent (i.e., 1 in 400) of human births. About 33% of identical twins have two complete and separate chorions, indicating that separation occurred before the formation of the trophoblast tissue at day 5 (Figure 1.11A). The remaining two-thirds of identical twins share a common chorion, suggesting that the split occurred within the inner cell mass after the trophoblast formed.

By day 9, the human embryo has completed the construction of another extraembryonic layer, the lining of the amnion. This tissue forms the **amnionic sac** (or water sac), which surrounds the embryo with amnionic

FIGURE 1.10   Twinning in humans. (A) Dizygotic (two-egg) twins are formed by two separate fertilization events. Each embryo implants separately into the uterus, and the resulting twins are no more closely related genetically than any two full siblings. (B) Monozygotic (one-egg) twins are formed from a single fertilization event. Sometime before day 14, the embryo splits in such a way that the cells of the inner cell mass are separated into two groups. Each ICM group forms a complete fetus, resulting in two individuals with identical genomes.

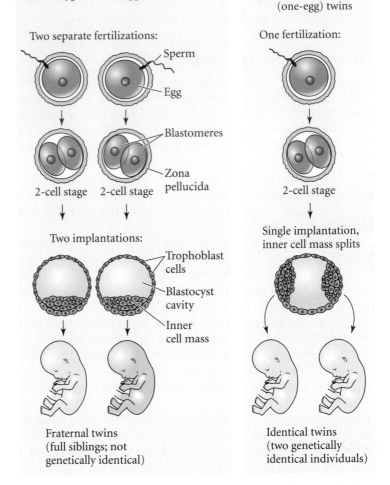

(A) Dizygotic (two-egg) twins

Two separate fertilizations:

Sperm

Egg

Blastomeres

Zona pellucida

2-cell stage   2-cell stage

Two implantations:

Trophoblast cells

Blastocyst cavity

Inner cell mass

Fraternal twins (full siblings; not genetically identical)

(B) Monozygotic (one-egg) twins

One fertilization:

2-cell stage

Single implantation, inner cell mass splits

Identical twins (two genetically identical individuals)

fluid that protects it from drying out and cushions it against impacts. If the separation of the embryo occurs after the formation of the chorion on day 5 but before the formation of the amnion on day 9, the resulting embryos have a single chorion but separate amnions (Figure 1.11B). A very small percentage of identical twins are born within a single chorion and amnion (Figure 1.11C). This means that the division of the embryo into two individuals came after the amnion formed at day 9. Such twins are at risk of being conjoined ("Siamese twins").

Twin formation thus demonstrates that the cells of the inner cell mass are undifferentiated and can become any part of the embryo. When undifferentiated cells from the inner cell mass are isolated and grown in certain con-

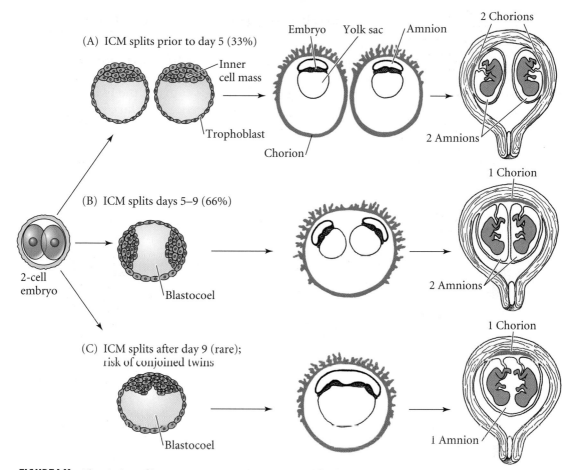

(A) ICM splits prior to day 5 (33%)

Inner cell mass

Trophoblast

Embryo   Yolk sac   Amnion

Chorion

2 Chorions

2 Amnions

(B) ICM splits days 5–9 (66%)

2-cell embryo

Blastocoel

2 Amnions

1 Chorion

(C) ICM splits after day 9 (rare); risk of conjoined twins

Blastocoel

1 Amnion

1 Chorion

FIGURE 1.11 The timing of human monozygotic twinning with relation to extraembry-onic tissues. (A) Splitting occurs before the formation of the trophoblast, so each twin has its own chorion and amnion; this occurs in approximately one-third of twin occur-rences. (B) In most of the remaining cases of identical twins, splitting occurs after tro-phoblast formation but before amnion formation, resulting in individual amnionic sacs but a single shared chorion. (C) Splitting after amnion formation leads to twins in one amnionic sac and a single chorion, a rare situation where there is a risk of conjoined ("Siamese") twins.

trolled laboratory conditions, they remain undifferentiated and continue to divide in their flasks. Such cells are called **embryonic stem cells (ES cells)**. As we will detail in Chapter 9, embryonic stem cells can be directed to form the precursor cells that give rise to blood cells, nerve cells, and other cell types. This gives them enormous potential value for repairing damaged organs.

Escape from the zona pellucida and implantation into the uterus

While the blastocyst moves through the oviduct on its trek to the uterus, it grows within a protein coat called the **zona pellucida.** The proteins in this coat were essential for sperm binding during fertilization; once fertilization has occurred, the zona pellucida proteins prevent the blastocyst from adhering to the oviduct walls. If such adherence does take place, it results in an **ectopic,** or **tubal, pregnancy**—a dangerous condition, because a blastocyst implanted in the oviduct will grow, rupturing the oviduct and causing life-threatening hemorrhaging.

When the embryo does reach the uterus, some 5–6 days after fertilization, it releases an enzyme that digests a hole in the zona pellucida (Figure 1.12A). Once out, the blastocyst makes direct contact with the uterus. The **endometrial cells** lining the inside of the uterus "catch" the blastocyst on a mat that the endometrial cells secrete (Figure 1.12B). This mat contains a sticky concoction of proteins that bind specifically to other proteins that are present on the embryo's trophoblast, thus anchoring the embryo to the uterus. Once this "anchor" is in place, the trophoblast secretes another set of enzymes that digest the endometrial protein mat, enabling the blastocyst to bury itself within the uterine wall.

At this point a complicated dialogue between the trophoblast cells and the uterus begins. First the trophoblast cells "invade" the uterine tissue,

FIGURE 1.12   Implantation of the mammalian blastocyst. (A) Once in the uterus, a mammalian (mouse) blastocyst "hatches" from the zona pellucida, a protein coat that prevents the embryo from prematurely implanting in the oviduct before it reaches the uterus. (B) Initial implantation of a rhesus monkey blastocyst in the endometrial cells of the uterine lining. (A courtesy of E. Lacy; B from the Carnegie Institution of Washington, courtesy of Chester Reather.)

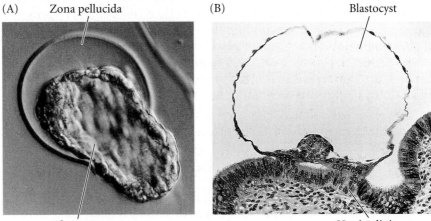

(A)   Zona pellucida

Blastocyst

(B)   Blastocyst

Uterine lining

secreting a hormone called human chorionic gonadotropin, or hCG. (This is the hormone that is measured in pregnancy tests.) Human chorionic gonadotropin then instructs the mother's ovaries to make another hormone, **progesterone**. Progesterone has a number of functions that are integral to the continued success of the embryo's development.* This hormone allows the uterus to remain malleable (i.e., soft and pliable, so that the embryo can grow); it also prevents menstruation (which would destroy the embryo) by blocking muscle contraction; and finally, progesterone allows the blood vessels from the uterus to surround the embryo.

Under the influence of progesterone, the uterus makes new blood vessels and starts to form a new region, the **decidua**. The decidua will become the maternal portion of the placenta. The decidual region then tells the trophoblast cells of the embryo to become the chorion. Together, the decidua and chorion form the placenta (Figure 1.13). Thus, the placenta is a single organ formed from two different organisms, the embryo and the mother.

Gastrulation

Gastrulation refers to the series of cell movements by which the embryonic cells change their positions relative to one another. The embryo at this stage is called a **gastrula**. Gastrulation begins about day 14 after fertilization (right around the time of the woman's first missed period). It is during gastrulation that the embryonic cells lose their ability to be pluripotent. They can no longer regulate—i.e., they cannot regenerate missing parts if some region of the embryo is removed. Thus, at gastrulation, the embryo is committed to become a single organism. It can no longer give rise to twins or other multiple births. This point is sometimes called "individuation."

> It is not birth, marriage, or death, but gastrulation, which is truly the most important time in your life.
>
> LEWIS WOLPERT (1986)

Early during the gastrula stage, some cells are set aside to become the **germ cells**—the precursors of the sperm or eggs. The rest of the cells that will form the embryo begin to interact with one another to determine their fates among the three major cell lineages. These lineages make up three **germ layers** (Latin, *germen*, to bud or sprout), each of which will "germinate" into the different tissue and organ systems (Figure 1.14):

- The **ectoderm** is the outermost layer of the embryo. It generates the surface layer (epidermis) of the skin, and also forms the brain and nervous system.

- The **endoderm** is the innermost layer of the embryo. It will give rise to the lining of the digestive tube and its associated organs (including the lungs).

---

*Chemically blocking the binding of progesterone to the uterine tissue prevents the implantation of the embryo and terminates the pregnancy. This blockade is the method of action by which the drug mifipristone—sometimes called RU486—produces an early abortion.

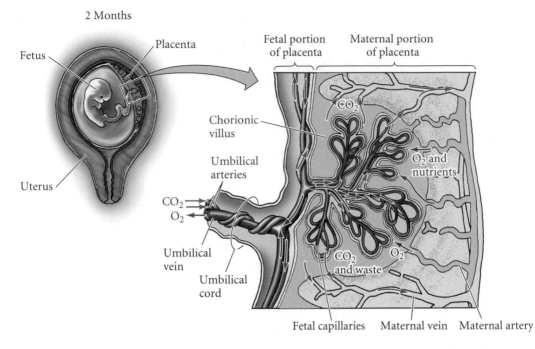

2 Months

Fetus

Placenta

Uterus

Fetal portion of placenta

Maternal portion of placenta

CO_2

Chorionic villus

O_2 and nutrients

Umbilical arteries

CO_2
O_2

Umbilical vein

CO_2 and waste

O_2

Umbilical cord

Fetal capillaries   Maternal vein   Maternal artery

FIGURE 1.13   The human placenta. Nutrients, oxygen, and waste materials (carbon dioxide and urea) are exchanged between the maternal and fetal blood in the placenta. The placenta forms from the chorion (trophoblast) of the embryo and from the decidua (uterine lining) of the mother. The embryo is attached to the placenta by its umbilical cord. Embryonic blood vessels enter the placenta and come into close proximity to the mother's blood vessels.

- The **mesoderm** is sandwiched between the ectoderm and endoderm. It generates the blood, heart, kidneys, gonads, bones, muscles, and connective tissues (e.g., ligaments and cartilage). Some of the mesoderm extends outward from the embryo and into the trophoblast. These cells will generate the blood vessels of the umbilical cord that connects the embryo to the placenta (see Figure 1.13).

In addition to the germ-layer cells that will form the embryo, the cells of the inner cell mass also form three sacs (see Figure 1.9): the amnion, the allantois, and the yolk sac. The function of the amnion was discussed earlier in this chapter. The **allantois** is a waste storage sac for the by-products of metabolism. The early embryo cannot urinate or defecate; its waste products are stored in the allantois until the placenta takes over these functions.

Unlike the prominent yolk sacs we associate with chicken eggs, the mammalian **yolk sac** does not in fact store any yolk. Because mammalian embryos

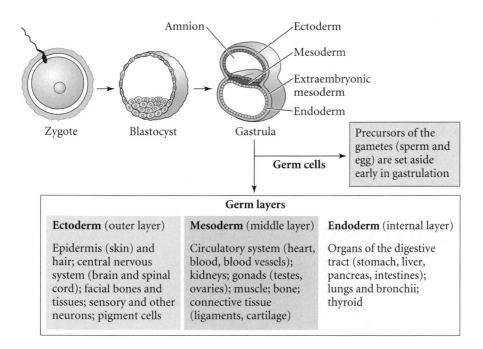

Germ layers		
Ectoderm (outer layer)	**Mesoderm** (middle layer)	**Endoderm** (internal layer)
Epidermis (skin) and hair; central nervous system (brain and spinal cord); facial bones and tissues; sensory and other neurons; pigment cells	Circulatory system (heart, blood, blood vessels); kidneys; gonads (testes, ovaries); muscle; bone; connective tissue (ligaments, cartilage)	Organs of the digestive tract (stomach, liver, pancreas, intestines); lungs and bronchii; thyroid

FIGURE 1.14   Gastrulation and the germ layers. As cells differentiate during the course of gastrulation, they are organized according to the germ layer from which they arise. Some organ systems contain cells from more than one germ layer. The germ cells (precursors of the sperm or egg) are set aside early in gastrulation and do not arise from the germ layers.

get nutrition through their placentas, our yolk sac is not needed for nutrition, as it is in bird and reptile eggs. The mammalian yolk sac, however, remains important because it is the place where the embryo's early blood cells are made, and where the germ cells (sperm and egg precursors) first form.

Organogenesis

Once the three germ layers are established, their cells interact with one another and rearrange themselves into tissues and organs. This process is called **organogenesis**. The first indication that organogenesis is occurring is the formation of the **neural tube** by ectodermal cells. The neural tube will become the brain and the spinal cord.

The ectodermal cells start as a flat sheet on the outside of the embryo (Figure 1.15A). At the start of neural tube formation, a ridge of ectodermal cells rises up on each side of the embryo and migrates toward the center (Figure 1.15B,C). As these ridges, or neural folds, migrate, they push the central cells

FIGURE 1.15   Neural tube formation is seen during the initial stages of organogenesis. The upper cells of the ectoderm—the neural ectoderm or neuroectoderm—fold inward , forming a tube. This neural tube will eventually become the central nervous system (CNS)—the brain and spinal cord. The sheet of ectoderm lying above the neural tube becomes the epidermis of the back, and the transient column of cells connecting the two is called the neural crest. The neural crest cells migrate to form the peripheral nervous system (that is, all the nerve cells that exist outside of the brain and spinal cord), the bones of the face, and several other cell types.

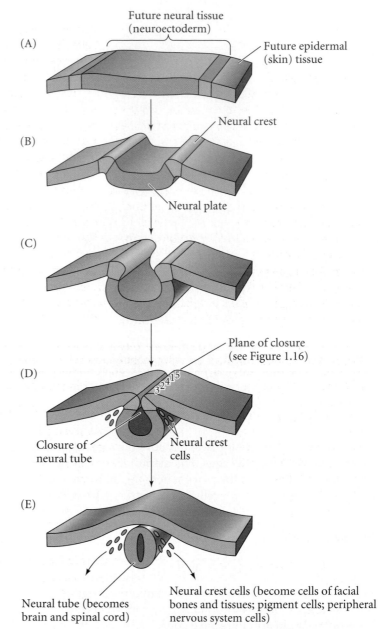

downward and then connect above them (Figure 1.15D,E). The cells that were at the center of the embryo are now the bottom (ventral or belly) cells of the neural tube, while those cells at the tips of the neural folds are now the top (dorsal, or back) cells of the neural tube.

Birth Defects

About 2 percent of human infants are born with some readily observable anatomical abnormality (Thorogood 1997). Those abnormalities seen at the time of birth are called **congenital anomalies** or, more commonly, **birth defects**. These abnormalities may include missing limbs, missing or extra digits, cleft palate, eyes that lack certain parts, hearts that lack valves, and so forth.

Abnormalities caused by genetic events (such as gene mutations or an abnormal number of chromosomes) are called **malformations**. Malformations often appear as **syndromes** (from the Greek, "running together"), where several abnormalities are seen concurrently. For instance, a malformation called Down syndrome is caused by an extra copy of chromosome 21 in each cell. People with an extra copy of this very small chromosome suffer from mental retardation, the absence of a nasal bone, heart defects, a characteristic slanting of the eyes, and often the closure of the intestine.

Congenital anomalies can also be the result of the fetus being exposed to exogenous (outside) agents, including certain chemicals, viruses, radiation, or high fevers. These anomalies are called **disruptions**. The agents responsible for disruptions are called **teratogens** (Greek, meaning "monster-formers"), and the study of how environmental agents disrupt normal development is called **teratology**. The table gives a partial list of known teratogens.

The summer of 1962 brought two portentous events in the study of teratology. The first was the publication of the book *Silent Spring*, in which biologist Rachel Carson described how the insecticidal chemical DDT, in widespread use around the

Some agents thought to cause disruptions in human fetal development[a]

Drugs and chemicals	Ionizing radiation (X-rays)
Alcohol	**Hyperthermia (fever)**
Antithyroid agents (PTU)	
Bromine	**Infectious microorganisms**
Cortisone	Coxsackie virus
Diethylstilbesterol (DES)	Cytomegalovirus
Diphenylhydantoin	Herpes simplex
Gentamycin	Parvovirus
Heroin	Rubella (German measles)
Lead	Toxoplasma gondii (toxoplasmosis)
Methylmercury	Treponema pallidum (syphilis)
Penicillamine	
Retinoic acid (Isotretinoin, Accutane)	**Metabolic conditions in the mother**
Streptomycin	Autoimmune disease
Tetracycline	(including Rh incompatibility)
Thalidomide	Diabetes
Trimethadione	Dietary deficiencies, malnutrition
Valproic acid	Phenylketonuria
Warfarin	

Source: Adapted from Opitz 1991.

[a]This list includes known and possible teratogenic agents and is not exhaustive.

Birth Defects (continued)

world, was destroying bird eggs and preventing reproduction in several bird species. The second was the discovery that thalidomide, a sedative drug used to help alleviate nausea in pregnant women, could cause limb and ear abnormalities in the fetus. These two discoveries showed that the embryo was vulnerable to environmental insults.

Thalidomide is a powerful example of a teratogen. Thalidomide is an effective drug with very few side effects for adults, and it was prescribed (mostly in Europe) as a mild sedative. However, in the early 1960s, two scientists independently concluded that this drug was responsible for a dramatic increase in a previously rare syndrome of congenital anomalies. The most noticeable of these anomalies was phocomelia, a condition in which the long bones of the limbs are severely shortened (Figure A). Over 7,000 affected infants

were born to women who took the drug, and a woman need only have taken one tablet to produce children with all four limbs deformed. Other abnormalities induced by the ingestion of thalidomide included heart defects, absence of the external ears, and malformed intestines.

That outside agents could affect the developing fetus was underscored in 1964, when an epidemic of rubella (German measles) spread across America. Adults who contracted the disease experienced relatively mild symptoms, but over 20,000 fetuses infected by the rubella virus were born blind or deaf, or both. Many of these infants were also born with heart defects and/or mental retardation.

Teratogens do their damage during particular times, as the body is developing certain parts. This time is called the **period of susceptibility**. Thalidomide, for instance, was found to be teratogenic only during days 34–50 after the last menstruation (about 20 to 36 days after fertilization). From day 34 to day 38, no limb abnormalities are seen, but during this period, thalidomide can cause the absence or deficiency of ear components (Figure B). Malformations of upper limbs are seen before those of the lower limbs because the arms form slightly earlier than the legs during development.

It is during the embryonic period that most of the organ systems form, whereas the fetal period is generally one of growth and modeling. The maximum period of teratogen susceptibility is between weeks 3 to 8, since that is when most organs are forming. The nervous system,

(A) PHOCOMELIA, THE LACK OF PROPER LIMB DEVELOPMENT, was the most visible of the birth defects that occurred in many children whose mothers took the drug thalidomide during pregnancy. (Photograph © Deutsche Presse/Archive Photos.)

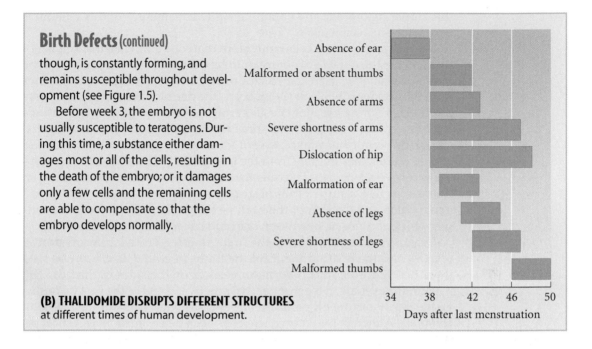

Birth Defects (continued)

though, is constantly forming, and remains susceptible throughout development (see Figure 1.5).

Before week 3, the embryo is not usually susceptible to teratogens. During this time, a substance either damages most or all of the cells, resulting in the death of the embryo; or it damages only a few cells and the remaining cells are able to compensate so that the embryo develops normally.

Absence of ear
Malformed or absent thumbs
Absence of arms
Severe shortness of arms
Dislocation of hip
Malformation of ear
Absence of legs
Severe shortness of legs
Malformed thumbs

34 38 42 46 50
Days after last menstruation

(B) THALIDOMIDE DISRUPTS DIFFERENT STRUCTURES at different times of human development.

In humans, this crucial formation of the neural tube begins early in the third week of gestation. The nervous system grows throughout embryonic and fetal development, and connections between the neurons begin to integrate the parts of the body and the brain. Different parts of the nervous system "zip" together starting from several points (Figure 1.16A). Failure to close the nervous system results in neural tube **birth defects** such as anencephaly (the anterior part of the neural tube remains open; Figure 1.16B) and spina bifida (the posterior end of the neural tube remains open; Figure 1.16C). Folic acid (also known as folate or vitamin B9) has been shown to be important in

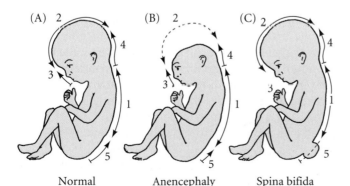

(A) Normal (B) Anencephaly (C) Spina bifida

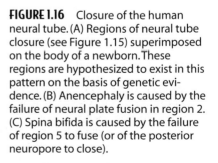

FIGURE 1.16 Closure of the human neural tube. (A) Regions of neural tube closure (see Figure 1.15) superimposed on the body of a newborn. These regions are hypothesized to exist in this pattern on the basis of genetic evidence. (B) Anencephaly is caused by the failure of neural plate fusion in region 2. (C) Spina bifida is caused by the failure of region 5 to fuse (or of the posterior neuropore to close).

this zippering process, and obtaining sufficient amounts of this vitamin is critical for pregnant women.

Many organs contain cells from more than one germ layer, and it is not unusual for the outside of an organ to be derived from one layer and the inside from another. For example, the outer layer of skin (epidermis) comes from the ectoderm, while the inner layer (the dermis) comes from the meso-derm. Also during organogenesis, certain cells undergo long migrations from their place of origin to their final location. The germ cells (primordial eggs and sperm), for instance, do not form within the gonads (ovaries and testes). Rather, they migrate from the base of the yolk sac into the develop-ing gonads—a long trip. The cells that form the facial bones migrate from the neural crest, a strip of cells originally seen between the newly formed neural tube and the skin of the back (see Figure 1.15E). Once these cells migrate, that connection between skin and neural tube vanishes.

Organs form rapidly during the first trimester. The heart forms during week 4, and the first rudiments of the limbs (legs and arms) can be seen then, too. The eyes start to form on week 5, on the sides of the head; by week 7, the eyes are in the front of the face. By the end of the first trimester, all the major anatomical parts are present, although many of them are not complete. Although the embryo's sex has been determined by its genotype, its sex organs are still rudimentary and "bipotential" (see Chapter 5). The sex organs don't become specifically male or female until the fetal period (around week 11).

The embryo at the end of first trimester is about 4 inches long and weighs about 1 ounce. By the end of second trimester, the fetus is about 12 inches long and weighs between 1 and 1.5 pounds. The nervous system continues to develop during the fetal period; the human electroencephalo-gram (EEG) pattern, a marker that the brain is functioning, is first detectable at around week 25 (that is, around the beginning of the seventh month of pregnancy).

Birth

The human baby is born as soon as its last critical organ system—the lungs—mature. If development were to go on too long inside the mother's body, the baby's head would grow bigger than the birth canal. However, if the fetus is born before the lungs mature, the baby cannot breathe on its own.

To coordinate the timing of birth with fetal lung development, humans have evolved an intricate system whereby the fetus can send a signal to the uterus, telling the uterus that it's time to start contractions. Once the fetal lung has matured, it secretes a type of protein called surfactant* into the

---

*Surfactant is a product of mature mammalian lung cells. Surfactant protein coats the inter-nal cells of the lung and keeps them moist in the presence of the air the lungs inhale.

amnionic fluid. This surfactant protein causes an immune response (similar to an inflammation) in certain fetal cells. These "inflamed" fetal immune cells migrate into the uterus, where they secrete proteins that in turn cause the uterine muscles to contract. In this way, a "protein cascade" from the fetus signals its mother that the lungs have matured and the baby is ready to "go it alone."

Like fertilization, birth is a process, and it is called (for good reason) **labor**. During the first stage of labor, contractions pull the cervix open and push the baby forward (Figure 1.17A–C). In the second stage of labor, the baby is born (Figure 1.17D,E). The third stage of labor pushes out the placenta (the "afterbirth"; Figure 1.17F), and the fourth stage is the recovery period.

When the baby takes its first breath, the air pressure closes a flap in the heart. This flap separates the blood circulation to the lungs from the blood circulation to the rest of the body. The baby can now breathe on its own, and the umbilical cord, which had been the source of oxygen for the growing fetus, can now be cut.

A human infant is born very immature. It cannot walk, and it cannot find its mother's breast without help. Even its eyes function only poorly. It is thought that this condition is brought about as an evolutionary compromise between the growing head and the size of a woman's pelvis. The human brain keeps growing throughout childhood, making millions of new nerve cells each day. If humans were born at the same stage of brain development as their ape relatives, a baby would probably be born at around 18 months, and its head would be far too large to pass through the birth canal. So it could be said that we spend the first few years of our lives as "extrauterine fetuses," totally dependent on parental care. We are still forming our nervous system during this time, and it has been hypothesized that our ability to think and interact with others is derived from the fact that we are being socialized during a time when we are rapidly forming neurons and can learn extremely rapidly.

Conclusion

This brief overview should have left you with at least the sense that the development of a human being comprises thousands of steps and finely coordinated interactions. At the cellular level, the embryo is a constantly growing and changing mass of cells that change from the single, totipotent zygote into a bewildering array of highly specialized cell types, including the many types of nerve cells that we blithely take for granted in our daily activities of working, playing, learning, and communicating. At the cellular level, thousands of proteins and smaller chemicals are involved in sculpting the human being from the masses of cells. A foreign

> The amazing thing about development is not that it sometimes goes wrong, but that it ever succeeds.
>
> VERONICA VAN HEYNINGEN (2000)

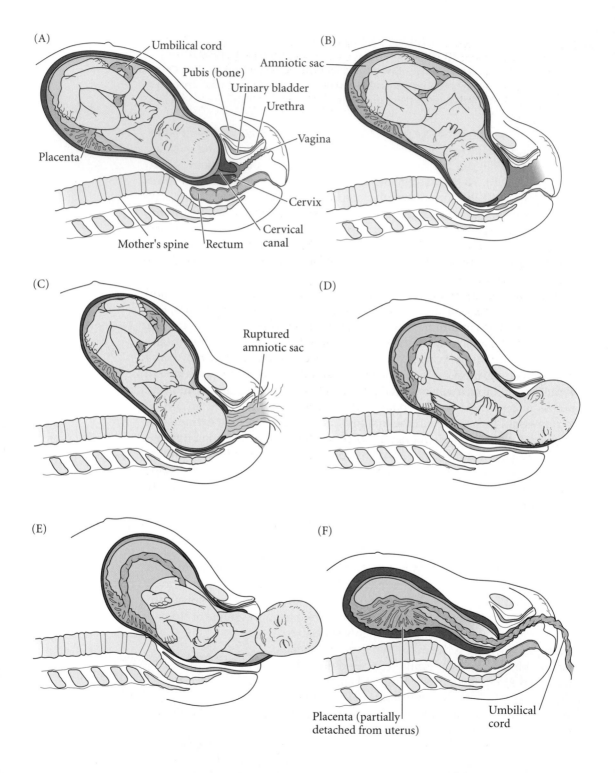

(A)
Umbilical cord
Pubis (bone)
Urinary bladder
Urethra
Vagina
Placenta
Cervix
Cervical
canal
Mother's spine   Rectum

(B)
Amniotic sac

(C)
Ruptured
amniotic sac

(D)

(E)

(F)
Placenta (partially
detached from uterus)
Umbilical
cord

◀◀ **FIGURE 1.17**   The major movements of labor initiated by contractions of the uterus. (A) Engagement, during which the head goes below the pelvic inlet. (B) Flexion, where the head enters the birth canal. (C) Descent and internal rotation during the first stages of labor. The cervix dilates (widens). (D) Extension of the fetal head ("crowning"). (E) External rotation of the body after the head is delivered. (F) After the baby is delivered, the placenta, or "afterbirth," separates from the uterus and will be expelled from the mother's body.

chemical getting in the way, or the lack of the right protein at the necessary time, can cause development to go tragically wrong. At the genetic level, the instructions in our nuclei cause us to look more like our biological parents than we look like our friends and neighbors.

Until recently, we could only marvel at the intricacies and constancy of human development. However, the incredible accumulation of molecular techniques and new knowledge in developmental biology has drastically changed both the science of developmental biology and the role of that science within society. Questions that were undreamed of outside of science fiction are now very real and are demanding answers every day. Finding answers, whether they apply to individual medical cases or affect society as a whole, demands not only an understanding of the science but also an appreciation and respect for the spiritual and philosophical ideals and foundations of human societies. The United States—faced not only with access to the most sophisticated technologies, but also with a population whose widely diverse religious, intellectual, and philosophical beliefs stem from virtually all the world's cultures—is especially embattled by these questions.

CHAPTER 2

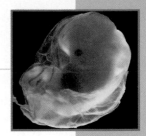

Philosophical, Theological, and Scientific Arguments

The question of when a human life begins underlies nearly all the other ethical questions considered in this book. People have thought about and asked this question throughout history. It has been expounded upon and argued by countless theologians, philosophers, doctors, lawyers, and politicians. Recent advances in science and medicine have led some people to look to biologists and other research scientists for answers—and their answers turn out to be as varied as those of any other group of thinking people.

The issue of life's starting point has become inextricably coupled with the issue of abortion. Like the question of when life begins, abortion has been a constant throughout human history. Whether a given society has condemned or condoned it, whether it has been legal or illegal, whether the procedure is medically safe or extremely dangerous, abortion has been a fact of human existence, and is likely to remain so. Today the moral acceptance of abortion extends from the question as to whether the procedure is manslaughter or simply the destruction of tissue.

"When does life begin?" may not be a new question, but modern scientific knowledge and recently developed new medical and technological abilities have made it an extremely prominent one. In this discussion, we will be looking specifically at individual "personhood." In the proper biological sense, "life" does not begin anew with each generation. The sperm and the egg are moving, metabolizing cells and are in fact biologically alive (although they will die unless they unite to form a zygote).

Cultural and Historical Perspectives

Prior to the twentieth century, it was not uncommon for newborn infants to die shortly after birth. Even in many contemporary societies, infant mortality remains extremely high. Thus, in many cultures, "personhood" is not accorded until a certain critical period after birth. In rural Japan, personhood is obtained when an infant utters its first cry. Among some tribes of northern Ghana, a child is said to acquire humanness 7 days after birth. For some Ayatal aborigines, personhood is not obtained until the child is named—and naming occurs 2 or 3 *years* after birth. For several Native American tribes in the Mojave Desert, human life begins for children who live long enough to be put to the mother's breast (Morowitz and Trefil 1992).

In reviewing some of the historical views that have shaped our sense of personhood, it is important to see them in the context of the scientific knowledge available *at the time*. Early human societies had no concept of the stages of development outlined in Chapter 1; the fact that a birth occurred approximately 288 days—or even 9 months—after an act of intercourse wasn't grasped by many, even in classical times (i.e., ancient Greece and Rome). Even when the male's role was understood, details were sketchy. Nobody knew sperm existed in the semen until the advent of microscopy in the 1600s. After the discovery of sperm, many people believed that each sperm contained a tiny, pre-formed human (a "homunculus"), and that the woman merely provided a nutrient-rich "soil" into which the seed (*sperma* is Greek for seed) was planted and grew.

It was not until the 1870s that scientists became certain conception was the union of the sperm and egg (see Pinto-Correia 1997). And it was not until the 1900s that the relationship between menstruation and the time of conception was fully understood. In fact, the absence of menses in a pregnant woman suggested to many that the embryo was formed from a woman's menstrual material, and that the man's semen somehow molded this material into an embryo.

Perspectives from history

The desire to raise strong males for battle led the ancient Spartans to frown upon abortion, although it was completely acceptable to most ancient Greeks to leave a sickly, weak, or deformed child on a hillside to die of exposure. Similarly, laws in the cities of the ancient Middle East discouraged abortion in favor of raising able-bodied men for the military. According to Assyrian laws, the penalty given a woman who induced her own abortion was impalement on stakes without burial (Rogerson 1985). It seems likely that the primary aim of these societies was not to protect unborn human life, but to ensure that the society did not lack military manpower (Buss 1967).

Among the ancient Greeks, the mathematician and philosopher Pythagoras (ca. 570–475 BCE) and his followers held a minority position in their disapproval of abortion (Tribe 1990). Although he is best remembered today for his studies in pure mathematics and geometry, Pythagoras led a colony dedicated to mathematics and the mystical union of the soul with the divine One. The Pythagoreans were vegetarians who believed in reincarnation and were opposed to both suicide and abortion.*

The wider Greek view was that of the philosopher Plato (ca. 472–347 BCE) who expounded the concept of **duality**: the idea that the soul and the body are separate entities. Plato appears to have been of the opinion that a human life begins at the point when a human soul enters the body, or the point of **ensoulment**. The concept of duality remains widely held today; however, the many differing opinions as to exactly when a human body becomes ensouled essentially can render it identical to the question "When does human life begin?" Plato appears to have believed that ensoulment occurs at birth.

The notion that ensoulment occurred at the time of birth was written into the laws of the Roman Republic. Although abortion was not openly endorsed by most Romans, it was not considered a serious offense (Buss 1967; Tribe 1990). In general, the Roman position was that the fetus was a part of the women's body during the duration of pregnancy and was ensouled only at birth. Indeed, in the years immediately following the birth of Jesus, the Roman philosopher Seneca (ca. 4 BCE–66 CE) disapprovingly stated that it was common practice for a woman to induce abortion in order to maintain the beauty of her figure.

The Aristotelian view

The most widely accepted and influential classical viewpoint was that of Plato's student Aristotle (384–322 BCE), who was both a scientist and a philosopher. Aristotle's position, which influenced Western thought (including that of the Catholic Church) for centuries, retained the tenet of the Spartans and other Greeks that deformed children should not be raised, although he objected to the practice of exposing healthy children merely for the purpose of population control (or because they were unwanted females;

---

*Hippocrates, the "Father of Medicine" (ca. 460–380 BCE), is sometimes portrayed as being against abortion, since the Hippocratic Oath expressly forbids giving a woman "an instrument to produce abortion." However, modern scholarship indicates that the Oath was not actually written by Hippocrates, but by a colony of Pythagoreans in the third century BCE (Edelstein 1967). Writings ascribed to Hippocrates himself do in fact mention means of abortion. Although a few modern medical schools still administer a form of the Hippocratic Oath to graduating medical students, virtually none uses the ancient version. As well as forbidding abortion, the original oath invokes the pagan gods, advocates the teaching of medical arts to men but not to women, and forbids physicians from "using the knife" (which was considered a skill separate from that of the physician).

see Chapter 6). He believed, rather, that the state should fix the number of children allowed to each couple, and that pregnancies in excess of this limit should be terminated before the point of "animation" (*Politics* 7:16, 1335b; see Bonner 1985).

According to Aristotle, animation was the point of ensoulment, the point at which an individual was created and subsequently possessed the form and rational power of a man (*History of Animals* Book 7, Part 3; see O'Donovan 1975). Because Aristotle believed in the physical and intellectual inferiority of females, and believed that this inferiority was due to their slower development in the womb, he concluded that animation occurred on the fortieth day after conception of male embryos and on the ninetieth day after conception of female embryos (Bonner 1985).

The importance of the "day 40" mark is that it is the time when the embryo begins to "look human" (see Figure 1.4). It is the time when the eyes come from the two sides to the front of the head, the external ears and nose become prominent, and the embryo assumes the "fetal position." Most Roman Catholic theologians prior to the nineteenth century—including St. Thomas Aquinas, St. Augustine of Hippo, and St. Jerome—espoused the view that the fetus is ensouled at around day 40.

English common law

The foundations of the modern legal systems of Great Britain and much of its Commonwealth, the United States, and Canada were laid in England in the middle ages. This body of "common law"—the source of, among many other things, the concept of trial by jury—has its origins in the twelfth-century reigns of Henry II and Richard Lionheart. Refined and expounded upon by English legal scholars over centuries, English common law has influenced the judicial systems of most Western societies.

English common law located the beginning of human life at **quickening**, the point at which the fetus can be felt moving within the uterus. Quickening occurs at approximately 120 days (4 months) of gestation. The 1973 decision of the U.S. Supreme Court in *Roe v. Wade* cited this precedent of common law in legalizing abortion in the United States.

Religious Viewpoints

Whether we see it as a strength or a weakness, moral certainty is the province of religion, not science. Certainly all the world's major religions have addressed, in some form or another, the question of when life begins. That there is disagreement among religious doctrines (as there is among scientists), and that religions have been known to change their views should be taken into consideration when one contemplates these questions.

Traditional Jewish views

The Jewish interpretation of when human life begins is extracted from three sources: the Torah, Talmudic law, and rabbinical writings.* Modern Judaism is far from monolithic, however, and includes a number of denominations that interpret the classical texts differently.

While the Torah does not directly discuss the beginning of human life or voluntary abortion, it does condemn miscarriage that results from violence toward a woman by an unrelated man. Exodus 21: 22–23 states that if a man injures a woman such that she survives but the fetus is lost, the perpetrator is penalized with a fine to compensate the family. If, however, the woman dies as a result of the violence, the attacker must "give life for life" and is executed, but no fine is incurred (Jakobovits 1973). This passage is usually interpreted to mean that killing a fetus is not equivalent to the murder of a human being and that human life does not begin during the fetal stage of development.

Talmudic law is more explicit in describing the point at which a fetus assumes personhood. At the point that the head of a full-term baby appears at birth (see Figure 1.17), the baby is awarded equal status to the mother's and can no longer be sacrificed to save her (Jakobovits 1973). Before this "crowning" of the head, the fetus has no legal rights as a human being. However, abortion is not usually permitted under traditional Jewish law, although it is considered acceptable if the mother's life is in danger. In this context, the fetus is viewed as being in "pursuit" of the mother's life, and may be destroyed as an "aggressor." Overall, then, the fetus is granted some recognition as a potential human life, but its status does not equal that of its mother's until its head can be seen coming out of the birth canal.

Although the Talmud gives the full status of humanness to a child at birth, some rabbinical writings have postponed the acquisition of humanness to the thirteenth day *after* birth for full-term infants (Jakobovits 1973). This designation is based on the viability of the infant, so the acquisition of humanness occurs even later for premature infants, because the viability of premature infants may still be questionable after 13 days (Buss 1967). (As mentioned earlier, many traditional cultures have a "waiting period" before a newborn is named or given legal status in the community, since without medical technology it is not uncommon for such infants to die.)

---

*The Torah, or "The Law," comprises the first five books (Genesis, Exodus, Leviticus, Numbers, Deuteronomy) of the Hebrew Bible (the Old Testament, or Tanakh). Traditionally believed to be the word of God as handed down to Moses, the Torah is the scriptural basis for Jewish law. The Talmud is rabbinical commentary and legal interpretations of the Torah and was originally an oral tradition (the "oral Torah"). The written Talmud was compiled by scholars from the second through the fifth centuries CE, and includes arguments and commentaries. More recent rabbinical writings are often included in Talmudic studies.

Early Christian views

The teachings of Jesus as articulated in the four Gospels do not specifically address the question of when life begins (although much is said about being "born again"). Likewise the apostle Paul, whose Epistles, along with the Gospels, are the foundation of Christian doctrine (the New Testament), has no definitive instruction on this point.

Early Christian interpretations of the Old Testament led to a distinction between an unformed and a formed (animated) fetus, the latter being considered an independent person with full human status (Buss 1967). This interpretation was embodied in the Justinian Code,* and branded as murderers those who caused the miscarriage of a formed fetus (Jakobovits 1973). The distinction between a formed and an unformed fetus raises the question as to how fully the writers of the Old Testament understood human development, and whether they designated a temporal period that marked the formation of a fetus. Passages in Job 10: 9–12 and Psalm 139: 13–16 hint at the process of fetal formation, though neither refers to a time frame in which it might occur (Rogerson 1985).

> Hast thou not poured me out as milk, and curdled me as cheese? Thou hast clothed me with skin and flesh, and hast fenced me with bones and sinews. Thou hast granted me life and favor, and thy visitation hath preserved my spirit.
>
> Job 10: 10–12

Christianity arose among the Jews and maintained a foundation of Hebrew scripture. However, the first century after the death of Jesus saw the new religion grow not so much in Hebrew Palestine, but in the classical Greek and Roman world. The apostle Paul, a Greek-speaking Jewish convert, traveled the Mediterranean preaching and writing epistles (letters) to other converts and would-be converts. By the time of his death around the year 67, Paul had succeeded in sowing the seeds of a religion no longer strictly bound by the laws of the Torah.

As Christianity gained converts in the classical world, there was increased friction between the rational, pragmatic philosophies of Graeco-Roman culture on the one hand, and the spiritual doctrines of Christianity on the other. In the second century, several Christian thinkers known as the Apologists,[†] or Defenders, worked to reconcile the teachings of Christianity with classical philosophy and to defend the new religion against rumors of

---

*The Emperor Justinian ruled from Constantinople (or Byzantium; present-day Istanbul, Turkey), the center of the largely Christian Eastern (Byzantine) Empire that grew in power after Rome fell to invasions by the Goths and other northern Europeans. The reign of the pious Justinian (527–565) was marked by the massive construction of public works and buildings (including many churches) and the codification of the Roman legal tradition into written volumes that still influence legal scholars.

†*Apology* is used here in its Greek root meaning of "a formal justification."

vile mystical practices such as human sacrifice (one such slander held that Christians killed and ate infants during the course of their communion rites). In 197, the Apologist Tertullian wrote denouncing both contraception and abortion, saying that as Christians,

> Murder being once for all forbidden, we may not destroy even the foetus in the womb ....To hinder a birth is merely a speedier murder; nor does it matter whether you take away a life that is born, or destroy one that is coming to the birth. That is a man which is going to be one; you have the fruit already in its seed.

This may be one of the earliest clear statements of the premise that life begins at the moment of conception. Tertullian did, however, recognize the need for abortions when necessary to save the life of the mother (Buss 1967; Bonner 1985).

In more recent times, some Christian theologians have argued that there is no point in distinguishing between a formed and an unformed fetus, because embryonic development is a divine process that should not be terminated by human intervention (Buss 1967; Rogerson 1985). Others argue that humanness is acquired on a continuum, and the state of humanness is reached through the acts of birth and baptism. It has been argued that, while degradation of a potential life should be avoided, true acquisition of humanness cannot be obtained until after birth, and miscarried fetal material is usually not accorded the rituals of baptism and burial with which many Christians note human birth and death (Rogerson 1985).

Positions of the Roman Catholic Church

During its history, the Roman Catholic Church has held varying positions on the beginning of human life. For most of the Church's history, its thinkers viewed immediate ensoulment at conception as impossible. Around 1140, when the monk Gratian compiled the authoritative canon law,* he concluded that "abortion was homicide only when the fetus was formed." Before the time of formation (about 40 days), the conceptus was not considered to be a fully ensouled human. Catholic doctrine as expounded by St. Thomas Aquinas (1225–1274) also followed the Aristotelian interpretation that a male fetus became ensouled at 40 days after conception, while the female fetus became ensouled at 90 days (Tribe 1990). Aquinas believed terminating a pregnancy prior to that time was sinful—a particu-

---

*Gratian's compilation, usually known as the *Decretum*, collected thousands of authoritative statements by popes, church councils, theologians, and secular authorities, to which Gratian added his own comments. The *Decretum* ("determined to be the case") quickly became the basic canon law textbook in the law schools of Europe and was a valid law book in the Catholic Church until 1917.

larly grave form of birth control—but was not abortion, *per se* (*Commentary on the Sentences III*, Dist. 3, Question 5; *Summa contra Gentiles II*, Chapter 89).

There were Catholic leaders who took exception to Aristotelian thinking. In 1588, Pope Sixtus V mandated that the penalty for abortion or contraception was excommunication from the Church; however, his successor, Pope Gregory IX, returned the Church to the view that abortion of an unformed embryo was not homicide. And in 1758, fear for the souls of those embryos that might die in the uterus caused Monsignor Francesco Cangiamila to publish *Embryologia Sacra*. This book advocated *in utero* baptism using a syringe—a practice that probably led to more than a few accidental abortions.

However, the Aristotelian view of ensoulment remained by and large the official view of the Roman Catholic Church until 1869, when Pope Pius IX again declared the punishment for abortion to be excommunication. Much of the support for his view was based on the idea that, since we cannot know with certainty the time at which human life begins, it should have protection from the earliest possible time, that of conception. Although it might not be ensouled, the fetus "is directed to the forming of men. Therefore its ejection is anticipated homicide." More recent Catholic theologians argued that the rational human soul in fact begins at the time of conception, because such an infusion must be a divine act. This argument has much earlier roots, having been put forth in 1620 by the physician Thomas Fienus, who claimed that the soul must be present immediately after conception in order to organize the material of the body (DeMarco 1984).

Today, Roman Catholic doctrine maintains the belief that animation or ensoulment is concurrent with the moment of conception. It also departs from the views of Tertullian and Augustine, who accepted the use of abortion when the mother's life was threatened. The modern Church asserts that "two deaths are better than one murder" (Jakobovits 1973). The Instruction *Donum Vitae* (1987) specifically states that "the human being is to be respected and treated as a person from the moment of conception; and therefore from this same moment his rights as a person must be recognized, among which in the first place is the inviolable right of every innocent being to life."

Some Protestant viewpoints

The many Protestant denominations of Christianity have taken widely divergent stands on issues such as slavery, homosexuality, and the admission of women into the clergy, so it is not surprising that there would be wide differences of opinion between and within Protestant congregations as to when human life begins. The Evangelical Lutheran Church in America is very open about these differences, acknowledging the diversity of viewpoints among its members. While recognizing that holding different views can be dangerous to the Church community, the Lutheran Church sees informed conversation about these issues as being beneficial, holding the possibility of clarifying one's beliefs concerning the roles of family and children, and concerning individual freedom and its limitations (www.elca.org/socialstatements/abortion/).

The Presbyterian Church of the United States accepts abortion as a last resort. Their stand appears more concerned with reforming the social environment than with worrying about when human life begins: "The Christian community must be concerned about and address the circumstances that bring a woman to consider abortion as the best available option. Poverty, unjust societal realities, sexism, racism, and inadequate supportive relationships may render a woman virtually powerless to choose freely" (www.pcusa.org/101/101-abortion.htm).

Some Protestant denominations claim authoritative knowledge of what the Bible dictates and will attempt to change laws in accordance with their beliefs. Thus, Resolution Number 7, "On Human Embryonic and Stem Cell Research," adopted at the Southern Baptist Convention in 1999 states that "The Bible teaches that human beings are made in the image and likeness of God (Genesis 1: 27, 9: 6) and protectable human life begins at fertilization." (www.johnstonsarchive.net/baptist/sbcabres.html).

As these three examples demonstrate, Protestant churches span the entire spectrum of positions on the beginnings of human life.

> Differences hold promise or peril. Our differences are deep and potentially divisive. However, they are also a gift that can lead us into constructive conversation about our faith and its implications for our life in the world.
>
> SOCIAL STATEMENTS OF THE
> EVANGELICAL LUTHERAN CHURCH IN AMERICA

Islamic views

The Islamic tradition has always placed great value on science and medicine. The works of the classical physicians were translated into Arabic, and the Arab world actively sought such scientific knowledge. By the time of the Prophet Muhammed (570–632), Arabic medical practices were the most advanced in the Western world. The Qur'an (Koran), transcribed by the Prophet during this time frame, largely reflects the thoughts of Aristotle and the Graeco-Roman physician Galen about the embryo (Musallam 1990).

Islam appears to espouse a view that strictly forbids abortion after the embryo has acquired a soul, something said to take place any time between 40 and 120 days

> Verily We created man from a product of wet earth;
>
> Then placed him as a drop in a safe lodging;
>
> Then fashioned We the drop a clot, then fashioned We the clot a little lump, then fashioned We the little lump bones, then clothed the bones with flesh, and then produced it as another creation. So blessed be Allah, the Best of creators!
>
> QUR'AN 23: 12–14

after conception (Tribe 1990). In 1964, the Grand Mufti of Jordan* declared that it is permissible to seek an abortion as long as the embryo is "unformed," which in his opinion was within 120 days of conception. Islamic law regards the fetus as a possible heir that can have its own heirs, but early abortion is only punishable when it is done without the father's consent (Buss 1967).

Eastern religious views

Hinduism, as practiced by millions in India, is a religion whose foundations are entrenched in the principle of *ahimsa*, or nonviolence. The practice of nonviolence is intrinsic to the Hindu belief in reincarnation—the repeated re-embodiment of the soul in different individuals and even different species. The *karma* (net cause-and-effect of one's choices and actions) generated in one's present life determines whether one's soul achieves a higher level or descends to a lower state in its next existence. The ultimate goal is to attain a state of bliss and enlightenment such that the soul is released from the cycle of earthly reincarnations and becomes one with Brahma, the Creator.

Hinduism teaches that abortion at any point is an act of violence, resulting in bad *karma* that will thwart the soul's progress toward enlightenment. Throughout the Vedas (the classical Hindu religious texts), pejorative references to abortion abound; it is referred to variously as "womb murder" and "the murder of an unborn soul" (see Tribe 1990).

The first of the five precepts of Buddhism is to avoid killing or harming any living being. Because the philosophy diametrically opposes the destruction of any form of life, even abortion to save the life of the mother violates the Buddhist ideal of self-sacrifice (for the mother). Its price is believed to be the woman's entrapment in the perpetual cycle of birth and rebirth (Tribe 1990).

Current Scientific Views

Science does not offer a hard-and-fast answer to the question of when human life begins, and there is no firm consensus among scientists' opinions. Indeed, the question "When does a human life begin?" tends to blur the lines between biology and social mores, since inherent within it is the idea of a "person," which is both a biological and a social category.

Biology might be better able to give information in answer to the question "When in the process of conception and gestation does an organism

---

*In Sunni Muslim countries, the Grand Mufti is the highest official of religious law. He prepares legal opinions, interpreting Islamic law for private clients or to assist judges in deciding cases. The Grand Mufti sometimes issues legal edicts, or *fatwa*. *Fatwa* are considered binding in civil matters such as divorce and inheritance, but are generally considered only as recommendations in criminal cases.

reach enough individualization to be considered a separate organism?" But even here, there is no clear-cut consensus, and some scientists argue that the process of becoming human is gradual, and that there is no specific point at which a nonhuman entity suddenly becomes human.

There are at least four stages of development that different scientists have claimed as the point where human life begins, including:

1. Fertilization (the acquisition of a novel genome)
2. Gastrulation (the acquisition of an individual physical identity)
3. EEG activation (the acquisition of the human-specific electroencephalogram, or brainwave, pattern)
4. The time of or surrounding birth (the acquisition of independent breathing and viability outside the mother's body)

Pictures and Words: Caveats for Sensitive Issues

Since most of us have neither seen an actual human embryo nor read a medical journal, our knowledge of these entities comes almost exclusively from photographs and magazine articles. These have to be approached with caution, and we offer three caveats to our readers.

The first caveat concerns the images of human embryos currently available in magazines, books, and websites. Several commentators have analyzed the ways in which human embryos and fetuses are often given autonomous status by dissecting them away from the mother and from the uterus, and even hiding the umbilical cord (see Petetchsky 1987; Franklin 1991; Taylor 1992; Stabile 1993; Gilbert and Howes-Mischel 2005). For instance, the cover of the June 9, 2003 *Newsweek* magazine bears the title, "Should a Fetus have Rights? How Science Is Changing the Debate." The cover image editorializes, however: its effect is to answer the question in the affirmative by showing a human fetus floating in space, wrapped only in its amnion. No uterus or mother is seen, and even the umbilical cord has somehow been removed. A person might easily believe this to be a scientifically correct image of the fetus, but it is not.

The second caveat concerns the language describing the embryos. One must remember that science is always done in the context of society, and this is especially true when studying something as socially sensitive as embryology and birth. Even the wording of human development differs from that of other animals. In humans, the term "embryo" refers specifically to the developing organism's first 8 weeks— the time of major organ formation (see Figure 1.4). After that, the organ systems have established their form, and the developing human is designated as a "fetus."

In recent years, there have been attempts to even further subdivide the terminology of human development. These reflect obvious social agendas. On the early end of the gestation period (roughly the first two weeks), one hears about the "pre-embryo." This defines the product of conception that has not individualized yet. It is a mass of cells that can still divide to form identical twins and other multiple siblings. Moreover, there is no certainty that a "pre-embryo" will develop into a fetus or newborn. At this stage of development, more than half of the products of conception die naturally before birth (see page 45). At the other

Pictures and Words: Caveats for Sensitive Issues (continued)

end of gestation (usually from weeks 28–38), one hears about the "unborn fetus." This is the stage when the organ systems are often sufficiently mature that, if born, the fetus would be likely to survive on its own (i.e., without modern medical support). All of these terms reflect societal and political distinctions placed on the biological processes of human development. They are neither needed nor used when discussing or studying the development of any other animal.

The language used in discussing abortion should neither ignore the value of unborn life nor the value of the woman and her other relationships.... The concern for both the life of the woman and the developing life in her womb expresses a common commitment to life. This requires that we move beyond the usual "pro-life" versus "pro-choice" language in discussing abortion.

SOCIAL STATEMENTS OF THE EVANGELICAL
LUTHERAN CHURCH IN AMERICA

The third caveat involves the language used to frame the debate. "Pro-life" is the belief that fertilization is the beginning of a human life; it has nothing at all to do with the length or quality of a person's postnatal life. (Indeed, it has often been pointed out that many people who describe themselves as "pro-life" are vehemently in favor of the death penalty.) "Pro-life" could as easily be called "zygote rights." Similarly, "partial birth abortion" is a highly charged conclusion about ending pregnancy that not everyone would agree with. The medical term for the procedure is "dilation and extraction."

In a similar vein, medical professionals and the medical literature refer to the natural termination of a pregnancy as a "spontaneous abortion." This accurate but (to some) offensive term is seldom heard outside of medical circles, where it is replaced by the word "miscarriage." The statement by the Evangelical Lutheran Church of America sums up well the need for sensitivity and understanding in choosing words that will not stop the possibility of rational debate before it starts.

View 1: You become human at fertilization

In this "genetic" view of human life, a new individual is created at fertilization (conception), when the genes from two parents combine to form a new genome with unique properties. This is a view that can be maintained with or without religious belief, and it is the position held by some scientists.

The "genetic" view is also the position of the Catholic Church, and it is the view held by many highly politicized anti-abortion activists. For instance, the anti-abortion website StandUpGirl (www.standupgirl.com) is very specific about when human life begins, telling us that "Fertilization marks the beginning of a new, individual life. At fertilization, the DNA of a single sperm and ovum merge to create the genetic blueprint for a new human being. Once the DNA has recombined and the single-celled ovum begins to divide, things really begin to roll."

One philosophical argument used to support the view that fertilization is when human life begins is the separation of "essential" from "accidental" characters. Thus, the religious philosopher Paul Ramsey quotes genetic evidence for the idea that being human is an *essential* property of the organism, and that it is defined by having a human genome. "Genetics," he writes, "teaches that we were from the very beginning what we essentially are in every cell and in every human attribute" (Ramsey 1970). Few scientists, however, would accept the notion that the genome determines "every human attribute." (This idea, termed "genetic determinism," is discussed in Chapter 14.)

Debates in the *New England Journal of Medicine* have focused on the issue of whether having a human genome is in fact the *sine qua non* of being a person. Anderson (2004) writes, "Does not the embryo possess all the genetic stuff of full humanness?" Sandel (2004) replies to this argument, writing that "the same thing can be said of a skin cell. And yet no one would argue that a skin cell is a person or that destroying it is tantamount to murder."

The entity created by fertilization is indeed a human embryo, and it has the potential to be a human adult. Whether these facts are enough to accord it personhood is a question influenced by opinion, philosophy, and theology rather than by science. Some scientists assert that the early embryo is not even an individual until it undergoes gastrulation.

View 2: You become human at gastrulation

This "embryologic" view proposes that a human receives individual identity around day 14, when the embryo undergoes gastrulation; it is at this point that the embryo can no longer form twins, and it is here that the cells begin the process of differentiation into the specific cell types of the new body. Because it is the point at which an embryo can give rise to only one person, many scientists consider gastrulation to be the point at which an embryo becomes an individual.

The embryologic view is expressed by scientists such as Renfree (1982) and Grobstein (1988), and has been endorsed theologically by Ford (1988), Shannon and Wolter (1990), and McCormick (1991), among others. Shannon and Wolter (1990) also raise the theological issue that, whatever ensoulment may be, it would not happen before day 12–14, since each twin is a distinctly different individual.

The view that a human does not become an individual before gastrulation, around day 14, is particularly crucial in the debate about allowing research on human embryonic stem (ES) cells, which will be covered in depth in Chapters 9 and 10. The embryologic view is consistent with the use of ES cells in biomedical research and has been supported as such by the conclusions of three national commissions: Britain's Warnock Committee (1984), the Canadian Royal Commission on New Reproductive Technologies (1993), and the NIH Human Embryo Research Panel in the United States (Parson 2004).

View 3: The acquisition of the human EEG pattern is when you become human

This "neurological" view of human life looks for symmetry between the ways we define human life and human death. Several countries (including the United States) have defined the end of human life as the loss of the cerebral EEG (electroencephalogram) pattern: death is determined by the "flatlining" of the EEG, even though the patient may have a heartbeat and be breathing. The "neurological" argument proposes that if the *loss* of the human EEG pattern determines the end of life, then its *acquisition* (which takes place at about 24–27 weeks) should be defined as when a human life begins (Morowitz and Trefil 1992).

Cerebral nerve cells accumulate in number and continually differentiate through the end of the second trimester of human pregnancy. However, it is not until the seventh month of gestation that a significant number of connections between the newly amassed neurons begin to take form. It is only after the neurons are linked via these *synaptic connections* that the wave pattern characteristic of active, conscious brain activity emerges. Just as a pile of unconnected microchips cannot function as a computer, the unconnected neurons of the fetal brain lack the capacity for conscious function prior to week 24. If one considers the quality of conscious awareness to define a human individual, this is a legitimate view of the starting point of a person's life.

View 4: You become human at or near birth

There is also the view that a fetus should be considered human when it can survive on its own. Traditionally, the natural limit of such viability was imposed by the respiratory system—a fetus could not survive outside the womb until its lungs were sufficiently mature, which occurs at about 28 weeks (see pages 26–27). Today, however, technological advances can enable an infant born as prematurely as 25 weeks to survive, although such infants are at high risk for having physical and/or mental disabilities.

Finally, there are those who believe human life begins when an individual has become fully independent of the mother, with its own functioning circulatory, alimentary, nervous, and respiratory systems. This traditional "birthday" is often recognized by seeing the head of the baby emerge or having the umbilical cord cut. One advantage of such moments is that they are well-defined, public, and obvious: the crowning of the head, the cutting of the umbilical cord, the first breath, or the first cry. In the absence of a clear consensus on when life begins, there are people who feel that birth is the only indisputable moment at which a conceptus becomes a human person (see Tooley 1973).

Biomedical research indicates that, even without including induced abortions, between 50 and 60 percent of all embryos conceived do not survive to birth (Mange and Mange 1999). Most of these embryos miscarry prior to the eighth week of pregnancy, and there is no assurance that any given egg, embryo, or fetus will survive to be born. John Opitz, a professor of pediatrics and human genetics, testified to the President's Council on Bioethics that our society is not prepared to value a fetus as person:

> If the embryo loss that accompanies natural procreation were the moral equivalent of infant death, then pregnancy would have to be regarded as a public health crisis of epidemic proportions: Alleviating natural embryo loss would be a more urgent moral cause than abortion, in vitro fertilization, and stem cell research combined.

In Conclusion

The question "When does life begin?" has been pondered by religious, secular, and scientific thinkers since the earliest recorded history. There have always been different opinions and answers, and even with immense knowledge of developmental biology now available, there is no consensus among scientists as to when human life begins. The stages of fertilization, gastrulation, brainwave acquisition, and independent viability each has its supporters. So does the view that there is no point at which one can say an embryo has suddenly become human, and that the whole question of "when does human life begin?" is framed in the religious perspective of "ensoulment" and thus cannot be answered scientifically. As the geneticist Theodosius Dobzhansky (1976) remarked:

> The wish felt by many people to pinpoint such a stage probably stems from the belief that a soul, conceived as preternatural entity, descends upon a formerly soulless living stuff, and suddenly transforms the latter into human estate. I hope that modern theologians can accept the idea that the transformation is not sudden, but gradual.

Any or all of these perspectives can be useful for contemplating what a human life is. Individuals will and should reach an answer that is meaningful for them, and most people indeed do so. However, in answering the question for themselves according to their own knowledge, experience, beliefs, and emotions, some people feel that they have also answered the question for everyone else. Such a mindset rejects the idea that there have always been diverse ways of thinking, that new information is constantly emerging, and that we are constantly reinterpreting our traditions as we integrate new knowledge and experiences into them.

UNIT 2

Should Assisted Reproductive Technologies Be Regulated?

Urge and urge and urge,
Always the procreant urge of the world.
Out of the dimness opposite equals advance,
Always substance and increase, always sex,
Always a knit of identity, always distinction,
Always a breed of life.

WALT WHITMAN (1855)

CHAPTER 3

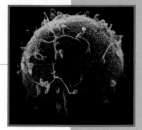

Fertilization and Assisted Reproduction

Fertilization is the process whereby the two sex cells—the **gametes**—fuse together to begin the creation of a new individual with genetic potentials derived from both parents. Fertilization accomplishes two separate ends: **sex** (the combining of genes derived from two parents) and **reproduction** (the creation of a new organism). Thus, the first function of fertilization is to transmit genes from parent to offspring, and the second is to initiate in the egg cytoplasm those reactions that permit development to proceed.

That fertilization is the union of sperm and egg is recently acquired knowledge. It was only in 1876 that Oscar Hertwig and Herman Fol independently demonstrated (in sea urchins) sperm entry into the egg and the union of the two cells' nuclei. Although the details of fertilization vary from species to species, it generally consists of four major events:

1. The sperm and egg must come into contact with each other and recognize each other as potential "partners": the cells must be gametes and members of the same species.
2. Only one sperm can ultimately fertilize the egg. Once a sperm has entered the egg, the egg prevents any further sperm from entering.
3. The genetic materials in the sperm and egg fuse.
4. A series of metabolic events is activated within the egg, and the events of development (see Chapter 1) begin.

In fertilization, both the sperm and the egg are active participants. The idea that the egg is the passive recipient of the male's sperm is a myth. As we will see, the human female reproductive tract activates the sperm, and the egg is probably involved in directing the movement of sperm toward it. The sperm can then activate the egg to divide and begin development.

Meiotic Cell Division

The critical difference between the gametes (sperm and egg) and somatic cells (the cells of the body) is the number of chromosomes in their nuclei. The mitotic cell divisions that generate the somatic cells allow the number of chromosomes in the nucleus to remain constant: each new human cell has 46 chromosomes and is genetically identical to every other cell in the body (see Figure 1.6). Each chromosome is represented twice (one copy each of 23 chromosomes from the mother, one copy each from the father). This normal condition is called the **diploid** nucleus.

In the cell divisions that generate the sex cells, however, *two successive divisions* reduce the number of chromosomes in the gamete nucleus to 23, called the **haploid** nucleus. Thus the egg and the sperm are ready to contribute 23 chromosomes each to the new individual's full complement of 46 chromosomes.

The type of cell division wherein the chromosome number is halved is called **meiosis** (Figure 3.1). After the DNA of each chromosome replicates, the homologous chromosomes (the two chromosomes 1, the two chromosomes 2, etc.) find one another and form a small cluster of four chromosomes, each pair held together by a **centromere**. The first meiotic division is very similar to mitosis, and one pair of chromosomes gets placed into each

FIGURE 3.1   Meiosis comprises two rounds of cell division. In the first division, the homologous chromosomes replicate their DNA and pair together. The two sets of chromosome pairs are separated from each other. During the second meiotic division, the centromeres split and the replicated portions of each chromosome separate. The result is a set of four haploid cells, each of which is unique (compare with the final panel of Figure 1.5). The diagram shows a single chromosome for simplicity; in an actual human cell undergoing meiosis all 46 chromosomes would be pairing and dividing simultaneously.

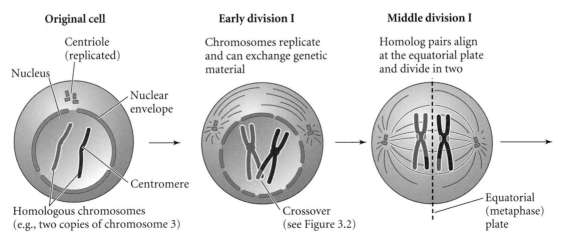

Original cell

Centriole (replicated)

Nucleus

Nuclear envelope

Centromere

Homologous chromosomes (e.g., two copies of chromosome 3)

Early division I

Chromosomes replicate and can exchange genetic material

Crossover (see Figure 3.2)

Middle division I

Homolog pairs align at the equatorial plate and divide in two

Equatorial (metaphase) plate

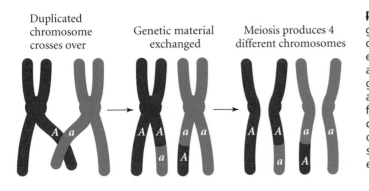

Duplicated chromosome crosses over

Genetic material exchanged

Meiosis produces 4 different chromosomes

FIGURE 3.2 Crossing over increases genetic diversity. An individual has two copies of each chromosome, one from each parent; here they are designated *A* and *a*. During meiosis, the arms of homologous replicated chromosomes can break and re-form (cross over), so that genes from chromosome *A* are transferred to chromosome *a*, and vice versa. Thus, some of the chromosomes resulting from meiosis will contain genes from both grandparents of the next generation.

daughter cell. During the second meiotic division, however, the centromere is split, and the "sister" chromosomes are separated into four new cells. The nucleus of each of these immature gametes is haploid (23 chromosomes).

There is another crucial difference between the products of meiosis and mitosis: while the two "daughter cells" resulting from mitosis are identical, each of the four sex cells produced by meiosis is genetically different. Each contains different combinations of chromosomes from the parental germ cells. One gamete might contain chromosome 1 from the father and chromosome 2 from the mother, chromosome 3 from the father, and so on. With 23 chromosomes from each parent, over 4 million combinations are possible. In addition, there exists within the processes of meiosis another, complex source of variation, **crossing over**—an exchange of chromosomal material that results in a "mixed" chromosome with genes from both parents (Figure 3.2). The phenomenon of crossing over can create virtually unlimited unique chromosomal combinations.

When the sperm and the egg combine their haploid nuclei, the diploid nucleus of 46 chromosomes is restored and development can begin. Since

Two daughter cells

Middle division II

In another round of cell division, the centromeres divide

Four genetically distinct haploid daughter cells

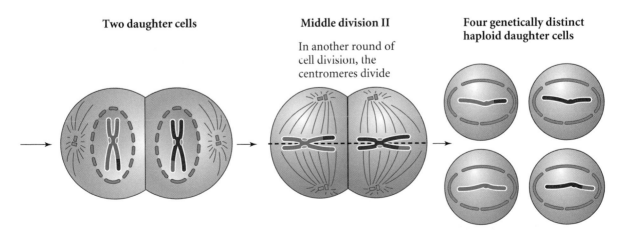

any sperm can combine with any egg, every human embryo starts off with a unique genetic constitution, or **genome**.

Structure of the Gametes

A complex dialogue exists between egg and sperm. The egg activates the sperm metabolism that is essential for fertilization, and the sperm reciprocates by activating the egg metabolism needed for the onset of development. But before we investigate these aspects of fertilization, we need to consider the structures of the sperm and egg—the two cell types specialized for fertilization.

Sperm

Each human sperm consists of a haploid nucleus, a propulsion system to move this nucleus, and a sac of digestive enzymes that enable the nucleus to enter the egg. Sperm and egg cells develop from germ cells (see Chapter 1); the diploid germ cells undergo meiotic division to become haploid gametes. The immature haploid sperm cell is called a **spermatocyte**.

Most of the sperm's cytoplasm is eliminated as the spermatocyte matures, leaving only those parts that are modified for the sperm cell's function (Figure 3.3). The haploid nucleus becomes very streamlined and its DNA becomes tightly compressed. In front of this compressed haploid

FIGURE 3.3   Mammalian sperm. (A) Diagram of a human sperm. The function of the cell is to deliver its nuclear material (blue) and centriole (yellow) into the egg; a mature sperm cell contains virtually no cytoplasm. The flagellum, using energy generated by the mitochondria, will propel the sperm to the egg. The proteins in the acrosomal vesicle will, when released, digest the membranes around the egg cytoplasm and allow the sperm to enter the egg. (B) A mature mammalian sperm cell from a bull. The DNA in the nucleus is stained blue; the mitochondria in the neck are stained green; and the microtubules of the flagellum are stained red. (Photograph courtesy of G. Schatten.)

(B)

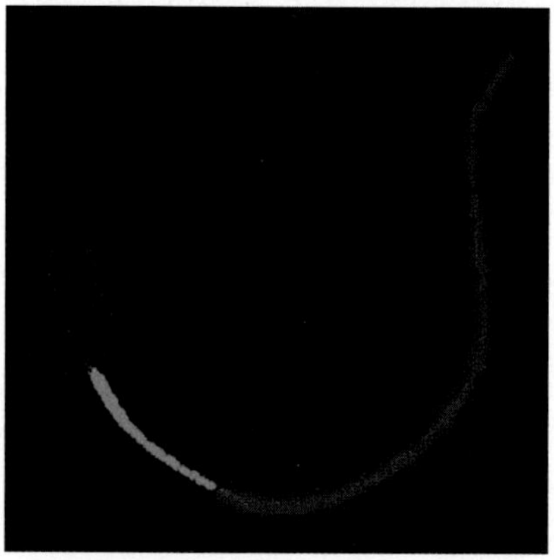

(A)   Sperm cell   Nucleus   Mitochondria
       membrane
       Centriole
    Acrosome

Head   Neck   Flagellum

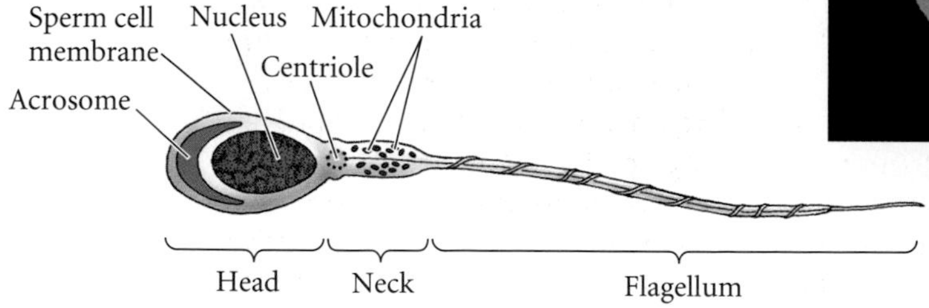

nucleus lies the **acrosome**, a membrane-enclosed sac containing enzymes that digest proteins and complex sugars. If a sperm manages to bind to an egg, these stored enzymes will be used to digest a hole through the outer coverings of the egg, allowing the sperm to enter the egg's cytoplasm and deliver its nucleus.

The human sperm is able to travel long distances thanks to the whipping motion of its taillike **flagellum**. Flagella are complex structures. The rapid movement of the flagellum is the result of microtubule proteins emanating from the centriole at the base of the sperm nucleus. These microtubules slide along one another and propel the sperm forward. The energy to move the microtubules comes from energy-generating structures called mitochondria (which can be found in many other types of cells as well) that are lodged in the "neck" of the sperm.

Thus, the sperm released during ejaculation have undergone extensive modifications to transport the nucleus to the egg. But even though they are able to move, these sperm do not yet have the capacity to enter and fertilize an egg. These final stages of sperm maturation, called **capacitation**, do not occur until the sperm has been inside the female reproductive tract for a certain period of time.

The egg

All the material necessary for the beginning of a new organism's growth and development must be stored in the mature egg, or **ovum**. Whereas the sperm cell eliminates most of its cytoplasm as it matures, the developing egg (called the **oocyte** before it is mature) conserves its material.

Unlike the process by which male germ cells become sperm, the meiotic divisions that form an oocyte are not equal. Rather, at each of the two meiotic divisions, one "daughter cell" conserves all of the cytoplasm, while the other product of the division is merely nucleus with a thin rim of cytoplasm. This structure, now called a **polar body**, will eventually degrade and disappear. So, although the sperm and the egg have equal haploid nuclear components (i.e., each contains 23 chromosomes), the egg also has a remarkable cytoplasmic storehouse wherein it has accumulated mitochondria, proteins, and protein-synthesizing molecules.

The egg's nucleus and cytoplasm are surrounded by several layers that the sperm must travel through (Figure 3.4). First the sperm must penetrate a "cloud" of cells called the **cumulus,** which is made up of the ovarian follicle cells that were nurturing the egg at the time of its release from the ovary. Next the egg has secreted the **zona pellucida**, a protein matrix that surrounds the egg and is critical in regulating sperm entry. (As detailed in Chapter 1, the zona pellucida also serves to regulate the time of embryo implantation should the egg be fertilized.) Finally, the sperm reaches the **cell membrane** that encloses the egg's nucleus and cytoplasm. This mem-

FIGURE 3.4   A mammalian egg immediately before fertilization. This unfertilized hamster egg, or ovum, is fairly typical of mammalian eggs. It is encased in the zona pellucida, which in turn is surrounded by the cells of the cumulus. A polar body, produced during meiosis, is visible within the zona pellucida. (Photograph courtesy of R. Yanagimachi.)

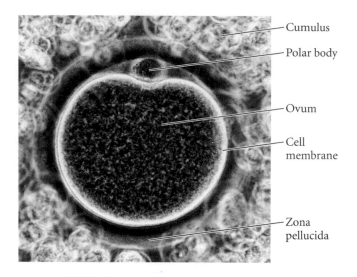

brane must regulate the flow of certain **ions**—electrically charged atoms—during fertilization and must be capable of fusing with the sperm's cell membrane.

Lying immediately beneath the egg cell membrane is a thin layer of gel-like cytoplasm called the **cortex**. The cytoplasm in this region is stiffer than the internal cytoplasm and contains membrane-enclosed sacs called **cortical granules** (Figure 3.5). The cortical granules are much like the acrosome of the sperm in that they contain protein-digesting enzymes.

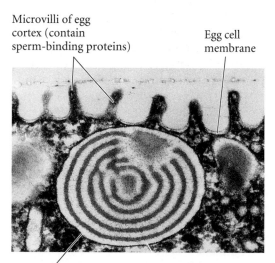

FIGURE 3.5   A cortical granule. This granule is from a sea urchin egg; human eggs contain similar structures. Cortical granules are found directly below the egg cell membrane. Once a single sperm has bound to the microvilli, the granules release enzymes that digest the sperm-binding protein, thus preventing any other sperm from entering the egg. (Photograph courtesy of T. E. Schroeder.)

They Meet: Recognition between Egg and Sperm

Generally speaking, there are six interactive steps by which the sperm and egg find each other and merge (Figure 3.6):

1. The egg and the female reproductive tract secrete proteins and biochemical molecules that attract and activate sperm cells.
2. A sperm cell binds to the zona pellucida of the egg.
3. The sperm's acrosome bursts, releasing its enzymes onto the zona pellucida.
4. The acrosomal enzymes digest of a path through the zona pellucida and the sperm passes through.
5. The egg and sperm cell membranes fuse.
6. The sperm (tail and all) enters the egg.

Sperm attraction and activation

Of the 280 million human sperm normally ejaculated into the vagina, only about 200 reach the **ampullary region** of the oviduct, where fertilization takes place. Although the internal fertilization process of mammals is difficult to study, we have learned that the female reproductive tract plays a very active role in the fertilization process.

Newly ejaculated mammalian sperm are unable to undergo the acrosomal reaction (the bursting of the acrosome and the release of its enzymes); they must first reside for some time in the female reproductive tract. As they pass through the uterus and oviduct, sperm cells interact with the cells and secretions of the female reproductive tract. These interactions result in physiological changes in the sperm that are critical for their ability to fertilize the egg. The set of physiological changes that allow sperm to become competent to fertilize the egg is called **capacitation.*** Sperm that are not capacitated are "caught up" in the cumulus and never reach the egg.

Interestingly, and in contrast to what is commonly believed about fertilization (and contrary to the opening scenes of the *Look Who's Talking* movies), the race is not always to the swiftest. Although some human sperm reach the ampullary region of the oviduct within half an hour after intercourse, those sperm may have little chance of fertilizing the egg. Nearly all human pregnancies result from sexual intercourse that takes place during a 6-day period ending on the day of ovulation. This means that the fertilizing sperm could have taken as long as 6 days to make the journey. Capacitation is probably a transient event, and sperm are given a relatively brief window of competence in which they can successfully fertilize the egg. As the sperm

---

*For purposes of in vitro fertilization (see pages 63–65), capacitation must be induced by incubating sperm in fluid from the oviducts, or in a manufactured fluid that contains calcium ions, bicarbonate, and serum albumin.

(A)

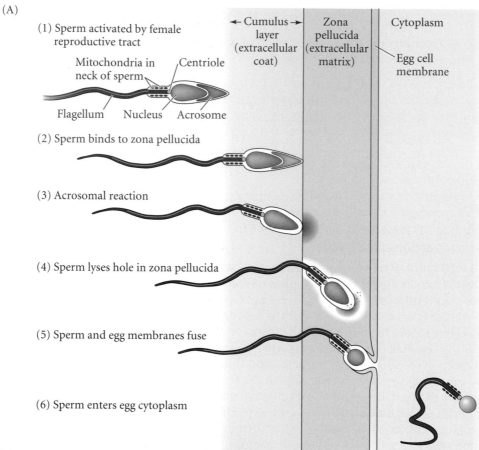

(1) Sperm activated by female reproductive tract

Mitochondria in neck of sperm   Centriole

Flagellum   Nucleus   Acrosome

← Cumulus → layer (extracellular coat)

Zona pellucida (extracellular matrix)

Cytoplasm

Egg cell membrane

(2) Sperm binds to zona pellucida

(3) Acrosomal reaction

(4) Sperm lyses hole in zona pellucida

(5) Sperm and egg membranes fuse

(6) Sperm enters egg cytoplasm

(B)

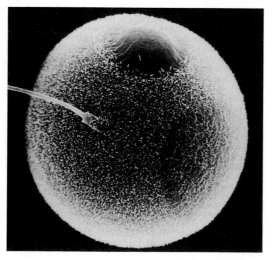

FIGURE 3.6   Summary of sperm-egg recognition and fusion events in mammalian fertilization. (1) The contents of the female reproductive tract capacitate, attract, and activate the sperm. (2) The acrosome-intact sperm binds to the zona pellucida (3) The acrosomal reaction occurs on the zona pellucida. (4) The acrosomal enzymes digest a path in the zona pellucida. (5) The sperm adheres to the egg, and their cell membranes fuse. (6) The sperm enters the egg cytoplasm. (B) Scanning electron micrograph of mammalian (hamster) sperm adhering to the zona pellucida. (Photograph courtesy of R. Yanagimachi.)

reach the ampullary region, they acquire competence, but if they stay around too long, they may lose it. Sperm may also have different survival rates (depending on their location within the reproductive tract), which could allow some sperm to arrive late but with better chance of success than those that arrived days earlier.

As sperm cells get closer to the egg, they become more mobile and move toward the egg. Recent studies have demonstrated the existence of proteins (and even temperature differences within the oviduct) that activate sperm motion and direct the sperm to the egg. The female reproductive tract, then, is not a passive conduit through which the sperm race, but a highly specialized set of tissues that regulates the timing of sperm capacitation and access to the egg.

Binding to the zona pellucida and cell fusion

The first stage of sperm-egg fusion takes place on the zona pellucida. Here, the sperm are bound by the proteins of the zona pellucida. The binding is critical, since these proteins activate the **acrosomal reaction**, in which the acrosome fuses with the sperm cell membrane and releases its contents (Figure 3.7). The acrosome's contents include enzymes that digest the zona pellucida and allow the sperm to wriggle its way to the egg cell membrane.

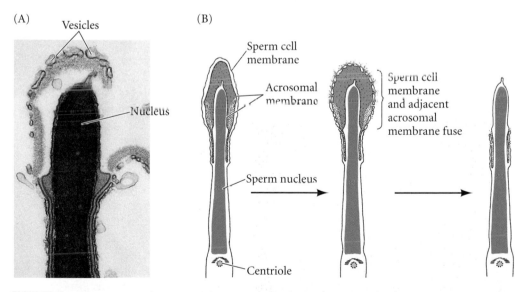

FIGURE 3.7   The acrosomal reaction. (A) Transmission electron micrograph of mammalian sperm undergoing the acrosomal reaction. The acrosomal membrane can be seen to form vesicles. (B) Interpretive diagram of electron micrographs showing the fusion of the acrosomal and sperm cell membranes in the sperm head, which releases the acrosomal enzymes that digest the egg's zona pellucida. (Photograph courtesy of S. Meizel.)

Contraception and Misinformation: Tragic Ironies

Much misinformation has been spread concerning contraception. In the United States, a group of some 1,500 pharmacists calling themselves "Pharmacists for Life International" has opposed filling prescriptions for contraceptive pills and for "emergency contraception." Emergency contraception, sometimes called "the morning-after pill," contains estrogenic hormones that prevent conception if taken within a few days of unprotected intercourse. One reason given by these pharmacists for their stand is that such drugs cause abortions (Jones 2004).

There is no scientific evidence that this is the case. The mechanisms of the "morning after pills" are presently being studied, but evidence suggests that these pills either stop ovulation or prevent sperm transport into the oviducts. In other words, they prevent the sperm and egg from ever meeting. Rivera and colleagues (1999) have studied the possible mechanisms by which these "morning-after pills" work and report that "no scientific evidence supports an abortifacient [abortion-inducing] effect."

There are about 3 million unintended pregnancies each year in the United States (Henshaw 1998). On average, a single act of unprotected intercourse occurring 1 to 2 days prior to a woman's ovulation is associated with an 8 percent chance of conception. Among women 19 to 26 years of age, however, the chance of conceiving can be as high as 50 percent. Using emergency contraception reduces this risk to 1–2 percent. (Wilcox et al. 1995; Dunston et al. 2002; Westhoff 2003).

Emergency contraceptive pills have been proven to be both effective and safe (Drazen et al. 2004). However, most women are unaware that such a pill even exists. Only 1 percent of women in the United States have used emergency contraception, whereas between 2 and 3 percent have had an induced abortion. In an ironic fashion, the pharmacists' refusal to fill prescriptions for emergency contraception (stopping sperm and egg from meeting) could result in many of those women having abortions (Grimes 1997).

Another piece of dangerous misinformation emanated from the Vatican and concerns condoms. In 2003, the president of the Vatican's Pontifical Council for the Family, Cardinal Alfonso Lopez Trujillo, made the statement that "the AIDS virus is roughly 450 times smaller than the spermatozoon. The spermatozoon can easily pass through the 'net' that is formed by the condoms … Relying on condoms is like betting on your own death." (quoted in Bradshaw 2003). The World Health Organization immediately denounced this view as being particularly dangerous.

World Health Organization studies as well as studies by the U.S. National Institutes of Health confirm that neither sperm nor the HIV virus can get through the rubber of an intact condom, and that proper condom use is extremely effective in preventing both pregnancy and AIDS (Bradshaw 2003). The WHO site documents that condom use in Thailand and Cambodia has resulted in an 80 percent drop in the transmission rate of HIV since the peak of the epidemics in those countries.

The late Pope John Paul II was adamant in the position that condoms could not be used even to prevent the transmission of AIDS. However, other prelates of the Catholic Church see it as the lesser of two evils. Kevin Dowling, a bishop from South Africa, has preached that opposition to condoms amounts to a death sentence for women, particularly in Africa, where cultural constraints render women unable to insist on either abstinence or fidelity (MacDonald 2004). In February of 2005, Cardinal Georges Cottier of the Vatican took the position that, while condoms should not be used

---

Contraception and Misinformation: Tragic Ironies (continued)

as contraceptives and are not the best way to stop the spread of HIV, the threat of AIDS is so immediate that "the use of condoms in some situations can be considered morally legitimate" (Arie 2005).

Cottier and Cardinal Juan Antonio Martinez Camino, a spokesman for the Spanish Bishops' Conference in Madrid, have said that one must balance two evils; it is no longer a question only of allowing the transmission of life, but of obeying the commandment "Thou shalt not kill." The tragic irony is that being "pro-life" regarding the creation of zygotes can make one complicit in the deaths of adult men and women.

When the sperm reaches this destination, the sperm and egg cell membranes fuse together. The sperm can now enter the egg cytoplasm.

The prevention of polyspermy

As soon as a single sperm nucleus enters the egg, the fusibility of the egg membrane, which was so necessary to get the sperm inside the egg, becomes a dangerous liability. Recall that each sperm that enters the egg can provide a haploid nucleus and a centriole to the egg. In normal **monospermy**, in which only one sperm enters the egg, a haploid sperm nucleus and a haploid egg nucleus combine to form the diploid nucleus of the fertilized egg (zygote), thus restoring the diploid chromosome number. The centriole, which is provided by the sperm, will divide to form the two poles of the mitotic spindle during cleavage.

The entrance of multiple sperm—**polyspermy**—leads to disastrous consequences. Fertilization by two sperm results in a "triploid" nucleus, in which each chromosome is represented three times rather than twice. Worse, since each sperm's centriole divides to form the two poles of a mitotic apparatus, instead of a bipolar mitotic spindle separating the chromosomes into two cells, the three sets of chromosomes may be divided into as many as four cells. Because there is no mechanism to ensure that each of the four cells receives the proper number and type of chromosomes, the chromosomes are apportioned unequally (Figure 3.8). Some cells receive extra copies of certain chromosomes, while other cells lack them. The result is an inviable zygote that cannot develop into a normal individual.

In humans, the block to polyspermy* involves the cortical granules of the egg (see Figure 3.5). When the sperm and egg membranes fuse, a signal is sent through the cytoplasm that causes the cortical granules to break open

---

*Blocks to polyspermy were discovered in the early 1900s by the African American embryologist Ernest E. Just, who became one of the few embryologists ever to be honored on a postage stamp.

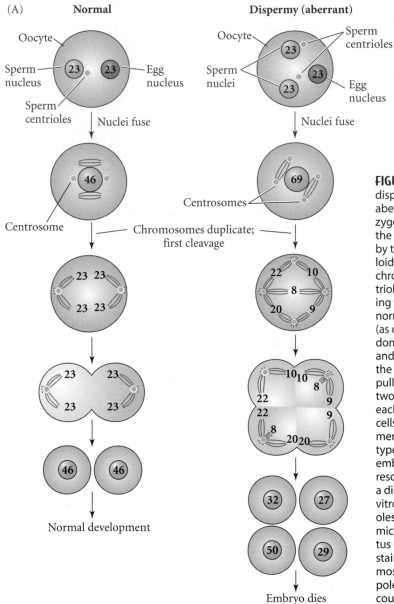

FIGURE 3.8   Consequences of dispermy. (A) Normal (left) and aberrant (right) development in a zygote. In the aberrant situation on the right, the egg has been fertilized by two sperm. There are three haploid nuclei (each containing 23 chromosomes) and two sperm centrioles (both of which divide, forming four mitotic poles instead of the normal two). The 69 chromosomes (as opposed to the normal 46) randomly assort on the four spindles, and during the first mitotic division, the chromosomes duplicate and are pulled to the four poles. Instead of two cells with 46 chromosomes each, the abnormal result is four cells containing a random assortment of different numbers and types of chromosomes; such an embryo dies very early. (B) The fluorescently stained micrograph shows a dispermic human egg (fertilized in vitro) at first mitosis. The four centrioles are stained yellow, while the microtubules of the spindle apparatus (and of the two sperm tails) are stained red. The three sets of chromosomes divided by these four poles are stained blue. (Photograph courtesy of G. Schatten.)

and release their contents into the space between the egg cell membrane and the zona pellucida. Like the sperm's acrosome, the cortical granules contain protein-digesting enzymes. The cortical granule enzymes digest the sperm-binding proteins on the zona pellucida, which prevents other sperm from

FIGURE 3.9   Calcium ions activate the egg. The entry of a sperm cell into the egg cytoplasm triggers a wave of calcium ions (stained with a fluorescent red dye in this starfish egg). The photos were taken at 5-second intervals, showing how the calcium ions spread through the egg, sending a signal that activates proteins and allows mitosis (cleavage) to begin. (Photos by S. A. Stricker.)

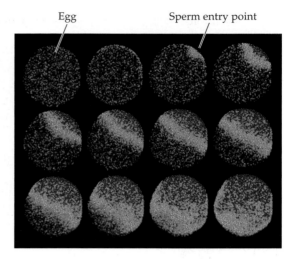

Egg   Sperm entry point

binding to the zona, as well as releasing any sperm that were already bound. The net result of these events is that, once one sperm has entered the egg, no other sperm can get inside.

Egg activation

In addition to blocking polyspermy, the fusion of the egg and sperm membranes also appears to cause the reactions that activate the metabolism of the egg and allow it to begin development. If the egg is not activated within a short time after it is ovulated, it will die. However, if it fuses with a sperm, a series of biochemical reactions are initiated that will result in development of the zygote. The key to this process appears to be the release of calcium ions that have been stored inside membranes within the egg. The fusion of egg and sperm cell membranes releases these calcium ions and they spread throughout the egg (Figure 3.9). These calcium ions activate protein synthesis and DNA replication, meaning that cell division (cleavage) can now take place.

Fusion of the Genetic Material

After the sperm and egg plasma membranes fuse, the sperm nucleus and its centriole separate from the mitochondria and the flagellum. The sperm mitochondria and the flagellum disintegrate inside the egg, so very few, if any, sperm-derived mitochondria are found in developing or adult organisms. Thus, although each gamete contributes a haploid genome to the zygote, the mitochondria are transmitted solely by the mother. Conversely, in almost all animals that have been studied (the mouse is the major exception), the centrioles needed to produce the mitotic spindle needed for the cell divisions of cleavage are derived from the father (i.e., the sperm's centriole).

> The elements that unite are single cells, each on the point of death; but by their union a rejuvenated individual is formed, which constitutes a link in the eternal process of life.
>
> F. R. LILLIE (1919)

FIGURE 3.10   Formation and movements of the haploid nuclei during human fertilization. The protein that makes up microtubules has been stained green, while the DNA shows up in blue. (A) The mature unfertilized oocyte completes the first meiotic division, budding off a polar body. (B) As the sperm enters the oocyte, microtubules condense around it as the oocyte completes its second meiotic division (top of cell). (C) After 15 hours, the two haploid nuclei have come together, and the centriole splits to organize the mitotic poles. (D) The chromosomes from the sperm and egg combine on the metaphase plate and a mitotic spindle initiates the first mitotic division. (Photographs courtesy of G. Schatten.)

(A)   Polar body

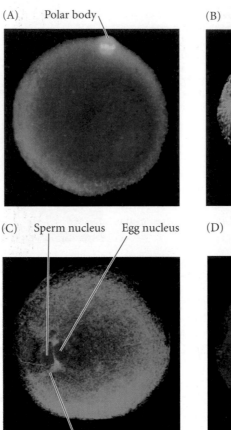

(B)   Sperm entry   Microtubules

(C)   Sperm nucleus   Egg nucleus

Microtubules

(D)   Sperm and egg chromosomes combine to initiate mitosis

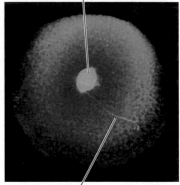

Sperm tail (disintegrating)

In mammals, the egg and sperm nuclei take about 12 hours to complete their migration and reach each other. The mammalian sperm enters almost tangentially to the surface of the egg rather than approaching it perpendicularly, and it fuses with numerous projections of the egg called microvilli. The mammalian sperm enters while the oocyte nucleus is "stalled" in the middle of its second meiotic division (Figure 3.10A); the egg cannot complete this division without the help of the sperm.* The second meiotic division is completed when the entry of the sperm causes calcium ions that have been stored in membranous sacs within the egg to be released into the

---

*Thus, one can say that the sperm isn't fully mature until it is capacitated near the egg, and the egg isn't fully mature until it incorporates a sperm. Each gamete completes the other's maturation process.

Fertilization: A Summary

Mammalian fertilization (which in humans is also known as conception) accomplishes two separate activities: sex (the combining of genes derived from two parents) and reproduction (the creation of a new organism).

The events of conception usually include (1) contact and recognition between sperm and egg; (2) regulation of sperm entry into the egg; (3) fusion of genetic material from the two gametes; and (4) activation of egg metabolism to start development.

The sperm head consists largely of a haploid nucleus and an acrosome. The nucleus contains the father's genetic material. The acrosome contains enzymes needed to digest through the zona pellucida surrounding the egg. The neck of the sperm contains a centriole that generates the microtubules of the flagellum, and mitochondria that generate the energy for flagellar motion.

The female gamete is the egg, which at the time of fertilization is technically in an immature state known as an oocyte. Its haploid nucleus contains the mother's genetic material. The egg contains an enlarged volume of cytoplasm, storing ribosomes (the protein-synthesizing units of the cell), mitochondria, and nutrient proteins. Cortical granules lie beneath the egg's plasma membrane.

The egg and the female reproductive tract secrete molecules that attract and activate the sperm. Sperm must be capacitated in the female reproductive tract before they are capable of fertilizing the egg. Sperm bind to the zona pellucida, and the proteins of the zona pellucida induce the acrosome of the sperm head to release its enzymes and digest a hole in the the zona pellucida. Fusion between sperm and egg cell membranes is probably mediated by protein molecules in both of the membranes.

Polyspermy results when two or more sperm fertilize the egg. Polyspermy is almost always lethal, since it results in at least three sets of chromosomes divided among four cells. Animals have evolved blocks to prevent more than one sperm from fertilizing an egg. In mammals, the sperm-binding proteins of the zona pellucida are digested once a single sperm enters the cytoplasm, so that no further sperm can bind (and any that were already bound will be released).

The fusion of egg and sperm membranes releases calcium ions from their storage areas in the egg cytoplasm. These calcium ions send a signal that triggers the completion of the egg's meiotic division, the synthesis of new DNA, and the resumption of protein synthesis. The zygote is now ready to proceed with cleavage and normal development.

egg's cytoplasm (see Figure 3.9). Once this division is completed and the egg has a haploid nucleus, the sperm and egg haploid nuclei migrate toward each other (Figure 3.10B,C).

As the nuclei migrate, DNA synthesis occurs in each haploid nucleus, and the centriole accompanying the male haploid nucleus produces the mitotic spindle. Upon meeting, the two nuclear envelopes break down (Figure 3.10C). The chromosomes orient themselves on a common mitotic spindle (Figure 3.10D) and proceed to complete the first mitotic division of cleavage. Thus, a true diploid nucleus in humans is first seen not in the zygote, but at the 2-cell stage.

The difference between male and female haploid nuclei

It is generally assumed that males and females carry equivalent haploid genomes. Indeed, one of the fundamental tenets of Mendelian genetics is that genes derived from the sperm are functionally equivalent to those derived from the egg. This is true for most genes. However, recent experiments in mice have shown that, in some cases, a particular gene will be active only if it came from the egg, while other genes are active only if they are brought by the sperm. Using laboratory mice, researchers have transplanted nuclei to create conceptuses with only sperm-derived or only egg-derived chromosomes. Neither case results in a functional embryo.

The same situation appears to exist in humans. In about one in every 1500 pregnancies, there is no embryo, but only a growth that resembles a mass of placental tissue. Such a growth is called a **hydatidiform mole**, and the majority of them arise from a haploid sperm fertilizing an egg in which the female nucleus is absent. After entering the egg, the sperm chromosomes duplicate themselves, thus restoring the diploid chromosome number—but the entire genome is derived from the sperm. The cells divide and have a normal chromosome number, but instead of forming an embryo, the egg becomes a mass of placenta-like cells.

Clearly, two chromosomal sets derived from only one parent (in this case, the male) cannot duplicate the effects of having a diploid genome provided by a male and a female parent. Genetic studies have shown that although most genes are the same whether they come from the sperm or the egg, a small group of genes will function only if they are from the sperm and a few other genes will function only if they come from the egg. Thus, a functional mammalian embryo can result only from the fusion of egg and sperm. (This will become important in our discussion of cloning in Chapter 7.)

Assisted Reproductive Technologies

Assisted reproductive technologies (ART) encompass several medical techniques that enhance the probability of fertilization by directly manipulating the oocyte outside of the woman's body. These techniques were designed to circumvent blocks to pregnancy and allow previously infertile couples to conceive a child. The techniques have evolved rapidly as medical science has tried to find ways to overcome all types of obstacles to bearing a healthy child.

In vitro fertilization

In vitro fertilization (IVF) is the oldest and most standard of the ART procedures. IVF involves retrieving eggs and sperm from the female and male partners and placing these gametes together in a petri dish* to enhance fertilization. After the fertilized eggs have begun dividing, the cleavage-stage embryos (see Chapter 1) are transferred into the female's uterus, where implantation and embryo development can occur as in a normal pregnancy. IVF was developed in the early 1970s to treat infertility caused by blocked or damaged oviducts in a woman. It is also used for couples when the man's sperm count is low, or if there are factors in his semen that interfere with conception. The first child successfully conceived using IVF was born in England in 1978. Since that time, the number of IVF procedures performed each year has increased and the success rate has improved significantly. If the woman is younger than 40 and her partner has no sperm problems, success rates for IVF compare favorably to natural pregnancy rates (that is, a woman's chance of becoming pregnant in any given month).

Because in vitro fertilization was the first assisted reproductive technology procedure to be developed and has been widely publicized, many people mistakenly believe that it is the only option for infertile couples. Actually, less than 3 percent of all patients who seek treatment for infertility will need to use IVF. Most infertile couples who seek evaluation and treatment respond well to less complicated treatment options, such as hormonal therapies (which cause more eggs to be ovulated) and artificial insemination (where sperm is injected into the uterus rather than having to travel through the cervix). The four-step IVF procedure described here and in Figure 3.11 remains the most commonly used assisted reproductive technology.

STEP ONE: OVARIAN STIMULATION AND MONITORING   Hormonal stimulation of the ovary causes several eggs (instead of just one) to mature at the same time. Physicians usually inject chemicals (artificial hormones) that instruct the woman's body to make more of the hormones that signal the egg maturation process within the ovaries. (Although they are usually called eggs, the female gametes that the physician will harvest are actually mature oocytes that are about to be ovulated.)

STEP TWO: GAMETE RETRIEVAL   Once so many eggs have matured that the ovarian follicles are close to rupturing, the physician collects as many eggs as possible using an aspiration pipette, a medical instrument that can "suck up" (aspirate) the eggs from the follicles. The two methods used to retrieve the eggs are ultrasound-guided aspiration, in which ultrasound images enable the physician to insert the pipette into each follicle to retrieve an egg; or laparoscopy, where the physician is guided by a visual image pro-

---

*"Test tube babies" are not conceived in test tubes, but in petri dishes. A petri dish is a shallow laboratory container traditionally made out of glass; hence the name *in vitro*, which is Latin for "in glass."

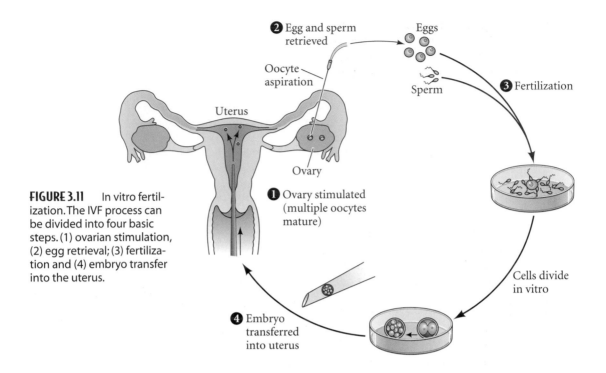

FIGURE 3.11 In vitro fertilization. The IVF process can be divided into four basic steps. (1) ovarian stimulation, (2) egg retrieval; (3) fertilization and (4) embryo transfer into the uterus.

duced by a narrow, flexible, lighted tube inserted into the abdomen. The recovered eggs are transferred to a sterile container to await fertilization in the laboratory.

A semen sample is collected from the male partner approximately 2 hours before the female partner's eggs are retrieved. If there is no evidence of low sperm number or semen deficiency, the man collects his own sperm through masturbation. If he has blocked sperm ducts or low sperm number, more invasive procedures can be used. These include surgically removing sperm directly from the testes or electrically shocking ejaculation when the nerves fail to function. The collected sperm are then processed using various laboratory techniques. Sperm processing (also called "sperm washing") helps the physician select only healthy and active sperm from the semen sample. Most commonly, the semen sample is placed into a thick solution and only those sperm that are able to swim to the top are collected. (This selects for motile sperm, and motility is usually a good sign that sperm are alive and well.) The selected sperm are then capacitated artificially by incubating them for a few hours in a solution containing high amounts of protein and calcium ions.

STEP THREE: FERTILIZATION Each of the mature, healthy eggs collected is placed in a petri dish with selected sperm, and they are incubated together at body temperature. In general, each egg is incubated for 12–18 hours with 50,000–100,000 motile sperm. Fertilization rates (as evidenced by the acrosome reaction, the blocks against polyspermy, and the formation of the second polar body) are between 50 and 70 percent. If all is successful, the fertilized eggs begin to divide, and embryos will shortly be ready for the next step: transfer into the uterus.

STEP FOUR: EMBRYO TRANSFER   Embryo transfer is not a complicated procedure and can be performed without anesthesia. It is usually done 3 days after egg retrieval and fertilization. This gives the physicians a chance to see which embryos are developing normally. A "good" embryo should have 6–8 cells of equal size, and there should be no evidence of cytoplasmic damage. Embryos from the petri dish are sucked into a catheter (a slender, tubular medical instrument). The physician then inserts the catheter through the female partner's vagina and cervix and inserts the embryos directly into the uterus. Normal implantation and maturation of the embryo is still required in order to achieve a successful pregnancy.

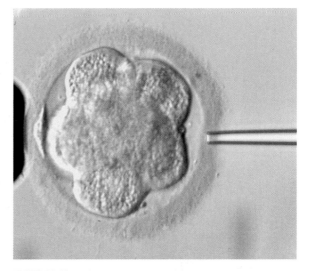

FIGURE 3.12   Assisted hatching. A hole is made into the zona pellucida to ensure that the embryo can hatch and implant into the uterus. (Courtesy of the Institute for Reproductive Medicine, Livingston, NJ.)

In instances where in vitro fertilization has been achieved but, after a number of cycles, implantation into the uterus consistently fails, the physician may suggest a procedure called assisted hatching. In assisted hatching, a microscopic hole is made in the zona pellucida prior to inserting the embryo into the uterus (Figure 3.12). This technique will usually ensure that the embryo is able to hatch from the zona pellucida (see Figure 1.12A) in time to adhere to the uterus.

Variations on IVF

In addition to the well-known sperm-meets-egg techniques of in vitro fertilization, other techniques are available for some couples.

Gamete intrafallopian transfer, or **GIFT**, was developed in 1984 as a variation of in vitro fertilization. The main difference between IVF and GIFT is that the harvested eggs and sperm are injected into the woman's oviduct, where fertilization will can take place naturally, rather than having fertilization take place in vitro and transferring the resulting embryos.

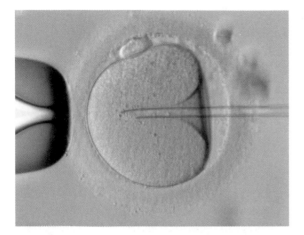

FIGURE 3.13   Intracytoplasmic sperm injection (ICSI). A single sperm is injected directly into the egg cytoplasm. The suction pipette holding the human oocyte is at the left; the injection pipette with the sperm inside it is on the right. (Courtesy of the Institute for Reproductive Medicine, Livingston, NJ.)

GIFT may be used in cases when the female partner has at least one unobstructed oviduct. GIFT is also recommended for patients whose infertility is due to cervical or immunological factors that prevent the sperm from reaching the oocyte in the oviduct. Although it is a more invasive procedure than traditional IVF, GIFT is used by some couples whose religion forbids fertilization in vitro.

Intracytoplasmic sperm injection (ICSI) was developed to treat couples who have a low probability of achieving normal fertilization due to the male partner's extremely low numbers of viable sperm (a condition called azoospermia). Male infertility may be due to genetic factors that lead to poor sperm production or to blockages or abnormalities of the ejaculatory ducts. It is also common for men who have had a severe injury to their reproductive organs, a vasectomy, or who have undergone chemotherapy or radiation for testicular cancer to have few sperm in their ejaculate.

In ICSI, a single sperm is injected directly into the cytoplasm of an egg (Figure 3.13). The fact that ICSI works has caused biologists to question their assumption that the only way for an egg to be activated is through sperm-egg membrane fusion. It now appears there are factors in the mammalian sperm head that are capable of activating the egg without interaction with the egg cell membrane.

Cryopreservation

The sperm used in IVF can be fresh, or it can be frozen and then thawed; the oocytes, however, cannot be frozen. Moreover, the 1- to 8-cell embryos generated by in vitro fertilization can be frozen and then implanted into the uterus. The first transfer of a **cryopreserved** ("frozen and thawed") human embryo came in 1983. In cryopreservion, the embryos are first incubated in dimethyl sulfoxide or some other agent to prevent ice crystals from forming and tearing the embryo. The embryos are then sealed in sterile glass containers and cooled to around −80 degrees Celsius. This cooling is also performed in a manner designed to prevent the formation of ice crystals. When the embryo is to be placed into the oviduct or uterus, this process is reversed, and the protective chemical is released from the embryo before it is inserted.

Conceptuses can be frozen for years—perhaps indefinitely. In general, embryos frozen at the zygote stage have a better chance of survival than those frozen at cleavage and blastocyst stages (Speroff and Fritz 2005). The implantation rate for cryogenically preserved embryos is about 10 percent—about one-half to one-third the success rate for IVF using fresh embryos.

Success rates and complications of IVF

The success rate of in vitro fertilization depends to a large degree on the age of the woman. Some recent statistics suggest that approximately 31 couples out of every 100 who try one round of gamete retrieval with IVF are likely to achieve pregnancy and delivery (Speroff and Fritz 2005). The success rate drops to 25.5 percent for women 35–37 years of age, and to 17.1 percent for women of ages 38–40. The Centers for Disease Control estimate that for woman over 40, the success rate drops to less than 5 percent, most likely due to the declining quality of eggs in women older than 40. (There are more abnormalities in meiosis as the egg cells age.) Compared to the 1 in 4 (25 percent) probability of achieving conception each cycle for normal healthy couples using unprotected intercourse, IVF is an optimistic treatment option for many infertile couples.

Although IVF is successful in achieving pregnancy in many women, it does carry the risk of multiple births (twins, triplets, etc.). It is believed that the transfer of more than one or two embryos increases the chance for pregnancy. However multiple embryo transfer also increases the risk of a multiple birth. The rate of multiple births depends on the age of the woman and the number of embryos transferred. When three embryos are transferred, the multiple birth rate is 46 percent for women age 20–29. The rate decreases to 39 percent for women age 40–44 when seven or more embryos are transferred. Thus younger women appear to be at greater risk for multiple births compared to older women, even when the number of embryos transferred is lower.

Multiple births are a serious concern because multiple-birth infants are predisposed to many health problems, including malformations, pre-term delivery, low birth weight, and death. Babies born prematurely and at low birth weight are at risk for cerebral palsy and chronic respiratory problems. In addition, women who carry multiple fetuses are at risk for many health conditions and complications, such as high blood pressure and diabetes.

Despite these drawbacks, artificial reproductive technologies are now widely practiced throughout the developed world. Some would argue that, at least in the United States, the technologies are too widely practiced, have too few safeguards to protect clients, and are available only to those who can afford them. The next chapter examines some of these arguments, and raises the question how (and if) such potent medical technology should be regulated.

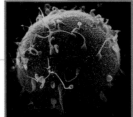

Assisted Reproductive Technologies: Safety and Ethical Issues

Assisted reproductive technology, or ART, was developed to allow infertile couples the chance to become pregnant. It has been extremely successful, in that the rates for deliveries following ART are just about equal to those for normal fertilization. However, the technology has raised many ethical and legal concerns about how safe these techniques are, who can make use of them, and how patients can be protected from fraudulent claims and practices.

Some religions still condemn ART; indeed, if one believes that a couple is infertile by the will of God, then ART would seem to circumvent the divine plan. But similar arguments have been used against many technologies. Smallpox vaccination was deemed immoral both because it undermined human dignity by injecting humans with cow pus, and because it prevented God's decreeing death upon those He thought deserving of it. Some clergymen deemed Benjamin Franklin's lightning rod immoral, since it circumvented the divine prerogative to blast whatever buildings or people He wanted destroyed in this dramatic way (see White 1898). Since it is not up to us to intuit what God's desire is for any given person, we will not consider this argument at length.

The Roman Catholic Church condemns assisted reproduction for the same reason it condemns birth control: it separates the physical act of marital union from the act of procreation. Pope Paul VI wrote in 1968 that there is an "inseparable connection, willed by God, and unable to be broken by man on his own initiative, between the two meanings of the conjugal act: the unitive meaning and the procreative meaning." This stand by the Church was reiterated in the 1987 proclamation *Donum Vitae*.

Regulation of Reproductive Technologies

In Great Britain, the site of the first "test tube baby" birth, strict laws regulate what a clinic offering ART services can or cannot do, and how they must report their results. In contrast, the United States government neither funds nor regulates research or treatment involving artificial reproductive techniques; it merely asks that each clinic gather its own statistics and report them to the Centers for Disease Control.

In the United States, there is a strong tradition and overwhelmingly held belief that the government should interfere as little as possible in sexual matters. There are laws forbidding incest, bigamy, and sexual relations with minors; aside from these, the U.S. government has largely avoided getting involved in reproductive issues. There have been no significant legislative moves to restrict procreation; in fact, persons with severe mental retardation are legally protected against compulsory sterilization or contraception.

In a 2004 review of the ART field, law professor John Robertson affirms that reproductive rights are considered individual rights, and as such are held against possible coercion by the state. However, in his view, reproductive rights do not extend to the right to have the state *provide* alternative reproductive services, and a private clinic may refuse or accept to offer services to anyone, as long as the clinic does not make these decisions on impermissible discriminatory grounds (Robertson 2004). Robertson calls these two principles "procreative liberty" and "provider autonomy," and he argues that these principles should frame any discussion on ART regulation. Clinics should not be restricted, he feels, since their purpose is to provide a way for potential parents to satisfy their desire to have children. Robertson concludes that more regulation is needed only if the harm from the technology is so great that it is "equivalent to the harm needed to limit coital reproduction."

> Legal restrictions would infringe the procreative liberty of infertile couples and require compelling justification.
>
> JOHN ROBERTSON (2004)

Taking an aggressive stance on the opposite side of Robertson's, attorney and bioethicist Lori Andrews argues that there is a pressing need for stringent regulation of ART. In her book *The Clone Age*, Andrews characterizes the current state of the field as a "Wild West" of medicine, where each woman undergoing treatment is an experiment (Andrews 1999). "In the course of my career," she relates, "I have learned several truisms: If it has worked in just one animal, it will be tried in a woman. If a baby is born from the technique, her picture will go up on the clinic wall, but no one will study how she fares as she develops, nor how her mother does over time." Andrews argues that without regulation, women are being treated experimentally and cannot know the risks of the procedures being used on them. She reports one infertility doctor as telling her, "We make things up. We try them on patients. And we

> A woman gets more regulatory oversight when she gets a tattoo than when she gets IVF.
>
> BROOKS A. KEEL, M.D. (1999)

never get informed consent because they just want us to make them pregnant" (Andrews 1999).

Should there be more oversight and regulation of fertility clinics in the United States? About 1 out of every 10 American couples is infertile (Mitchell 2002), and each year more than 1 million American women use infertility services. Many of these women are able to conceive using less invasive methods, such as artificial insemination or the stimulation of ovulation. However, about 100,000 pregnancy attempts are made each year using in vitro fertilization (IVF), resulting in the birth of about 30,000 babies. More than 177,000 babies have been conceived this way in the United States; throughout the world, it is estimated that over 1,250,000 children have been born as a direct result of IVF (Jones 2005).

Infertility services in the United States have grown into a 2-billion-dollar-a-year industry, and couples using IVF typically spend somewhere between $44,000 and $200,000 to achieve a single pregnancy (Andrews 1999; Caplan 2005). However, there is no standardization either of treatment procedures or of reporting results in American clinics. It is difficult to compare success records or health records between clinics. False advertising has been a big problem, with some clinics artificially inflating their success rates by including multiple deliveries or sometimes just lying. Thus, the major ethical issues (at least in the United States) focus on the very practical and intertwined concerns involving the safety of the IVF procedures and the regulation of IVF clinics.

Are ART Procedures Safe?

There are several points during the procedure where the techniques of in vitro fertilization can damage the mother or the conceptus. While these are known among the medical community, some critics have alleged that this information is not being released to the public. Questions have also arisen about the overall health of children conceived by IVF as compared to those whose conception occurred in the time-honored way.

Oocyte procurement

In order to harvest large numbers of eggs for IVF, the woman first has to take drugs or hormones that cause many oocytes to mature at once (instead of the usual single oocyte). Such hormone treatments can create problems with hypertension and mood swings.

Once the eggs are ready to harvest, the most prevalent surgical complication is minor hemorrhaging (loss of blood), which occurs in 8 percent of women undergoing IVF. More serious hemorrhaging is a relatively rare occurrence (around 1 in every 2,000 procedures). However, a recent study notes that "there is wide variation in the way this common procedure is performed, with room for improvement through published guidelines" (El-Shawarby et al. 2004).

Ovarian hyperstimulation syndrome—a combination of nausea, vomiting, and abdominal pain caused by increased fluid in the abdomen and chest—is also a medical concern for women undergoing IVF. The syndrome is caused by very enlarged ovaries and increased fluid triggered by an unexpectedly large response to the oocyte-maturing hormones given during IVF. Most patients have a mild form of this syndrome that begins about 4 days after egg collection and goes away within days. However, in 1–2 percent of cases, women develop severe ovarian hyperstimulation syndrome and can have kidney damage, ovary damage, and blood clotting disorders (IVF-infertility.com 2004).

Sperm surveillance

One would not think that it would be hard to keep track of the male's sperm once it had been obtained. However, at a Florida clinic, a white woman gave birth to children that were obviously not from her husband's sperm; her husband was black, but the babies were white. Blood tests confirmed that the clinic mistakenly used someone else's sperm for her IVF procedure. A similar instance occurred in the Netherlands, where a technician apparently did not clean the pipette between uses and the woman gave birth to twins of two different races (van Koij et al. 1997).

Multiple births

Probably the most pressing medical issue in IVF concerns the transplantation of several early embryos into the uterus or oviduct. Since in any given IVF attempt the chances of any one embryo coming to term and being born is much less than 100 percent, the chance of a woman having a successful pregnancy is made greater by inserting two or more embryos into the uterus. One consequence of this practice may be the twofold higher risk of tubal pregnancy (see page 18) for women who undergo IVF. This may seem strange, since the embryos are inserted into the uterus and not the oviduct; but the placement of the catheter and/or the high volume of fluid injected with the embryos can cause them to "back up" and implant in the oviduct instead of the uterus (Braude and Powell 2003).

Multiple embryo implantations are also the cause of multiple births. Multiple births may hold a particular fascination for the public, but they are a huge concern for physicians and hospitals. In 1997, when Bobbi McCaughey gave birth to septuplets, the newspapers called it a miracle, and she received enormous amounts of aid, money, and publicity. Many physicians probably thought it was closer to malpractice than a miracle. The McCaughey septuplets were lucky in that they survived at all. Even so, these babies had to be hospitalized for 3 months. The youngest was sent home with an oxygen tank, and soon returned to the hospital with breathing problems, while one of her brothers needed eye surgery.

In 2000, 35 percent of all births from IVF in the United States were multiples (CDC 2000). This is *10 times* the national average for multiple births. The high rates of multiple births associated with IVF has both financial and medical consequences.

In 2005, hospital charges for the delivery of a single infant were around $10,000. This cost rose to over $100,000 for twins. It has been estimated that by the year 2000, multiple births from ART had cost the American public some $640 million in medical expenses surrounding the birth of the infants (and not including the significant costs for the care of long-term complications). These high costs are due to the infants' being born prematurely and at low birth weights. Multiple-birth children are much more likely than singles to be born prematurely, and to have much longer and more intensive hospital stays. Only 15 percent of single-born babies require neonatal intensive care, whereas 78 percent of multiple-birth children do.

Compared to single births, multiple-birth infants have a higher death rate and a higher risk of long-term physical and mental handicaps, including cerebral palsy, developmental delays, and blindness. In 1 out of every 5 triplet pregnancies and 1 in 2 quaduplet pregnancies, at least one child has a severe long-term handicap (American College of Obstetricians and Gynecologists 2004). Because of these risks, previously infertile couples may be faced with the ironic and often emotionally painful choice of "selective fetal reduction": the abortion of one or more fetuses in order to increase the chance of carrying at least one fetus to viability, and to reduce the risk of the child's long-term physical impairment.

In England, the law stipulates that physicians may implant no more than 3 embryos at a time. There are no such laws in the United States, and some physicians routinely transplant 7–10 embryos. Indeed, reproductive clinics often transfer large numbers of embryos not only because the mother desperately desires to become pregnant but also because it makes the clinic's statistics (number of babies produced by IVF, number of babies per cycle, etc.) look much better (Andrews 1999; Jain et al. 2002).

Guidelines published by the American Society for Reproductive Medicine and the Society for Assisted Reproductive Technology (summarized in Speroff and Fritz 2005) propose that no more than 2 embryos be transferred into women who are 37 years old or younger, and that 3–4 embryos are optimal in women between the ages of 38 and 40.

Offspring health

Another issue is whether or not IVF procedures produce healthy adults, even if the baby is a single birth. This issue is very controversial, since IVF is a recent procedure and relatively few IVF offspring have reached adulthood yet. However, animal studies show that preimplantation maternal diet can be crucial for adult health (Kwong et al. 2000), and the nutritional adequacy of IVF culture medium for adult health is not known (McEvoy et al. 2000).

While IVF appears to double the risk of a single-born baby having low birth weight or a major birth defect, this might not be a major worry, since even accounting for this increased risk the likelihood of a full-term baby of normal birth weight is about 94 percent and the likelihood of having an infant free from major birth defects is 91 percent (Mitchell 2002).

There has also been controversy over intracytoplasmic sperm injection (ICSI; see page 68) because this procedure gets around the natural barriers intended to insure a sperm's health. However, one survey of the medical literature found no significant difference in the incidence of major birth defects between babies conceived naturally and those born from ICSI procedures (Lewis and Klonoff-Cohen 2005). Some studies showed a slight increase in chromosomal anomalies or birth defects, while other studies did not (Hansen et al. 2002; Winston and Hardy 2002). For whatever reason, however, the mental maturity of 1-year-old children conceived using ICSI lagged behind that of babies conceived in the normal way, although some studies show that the scores equalize by the age of 5 years.

Law professor John Robertson has argued that the issue of potential offspring health problems should not be a restrictive one, nor should it be a cause for regulation. His legal argument is that "restricting ART to protect offspring from the risks of ART runs afoul of Parfit's non-identity problem.* Once children have been born with those conditions, they would not have been wrongfully wronged or harmed, since there was no other way for them to have been born" (Robertson 2004).

Some Ethical Questions

Who benefits from ART?

Procedures such as IVF are provided as medical necessities for infertile couples in Denmark, whereas in the United States, couples can get such treatment only if their insurance covers it or they have the money to pay for it themselves. A few states require insurance companies to cover IVF, some states have voluntary coverage (i.e., coverage is left up to the individual insurance companies), and most states have no legislation at all.

---

*"Parfit's non-identity problem" is a legal argument based on the premise that an act can wrong a person only if the net result of the action is that the person is worse off than they otherwise would have been (Woodward 1986). In this instance, the argument states that a person who has a disability stemming from ART would not have existed at all without the intervention of ART.

Assisted reproductive procedures are expensive. The average cost per delivery in 1994 was $66,667 for couples undergoing their first cycle of oocyte maturation, rising to over $114,250 for those couples on their sixth attempt (Neumann et al 1994). This does not include the hospital fees for multiple births (which often exceed $100,000). Thus, the current availability of ART procedures in the U.S. certainly discriminates against those without resources. Some critics of this disparity suggest that each state require medical insurance to cover a certain number of IVF attempts for any citizen who wants them (see Jain et al. 2002).

Moreover, an infertility clinic can decide who it serves and who it turns away. A 2005 survey found that most U.S. clinics would help a 43-year-old woman get pregnant. One in 5 would refuse single women, but 5 percent said they didn't even ask about marital status. About 25 percent would help a woman who tested positive for HIV (the AIDS virus). A financially secure gay couple and a heterosexual couple on welfare were about equally likely to be turned away (Gurmankin et al. 2005). And each clinic has different criteria; as Caplan (2005) notes, one clinic might refuse you, but "just down the road is another place that might take you."

Is using ART procedures to allow 50-year-old women to have babies the best use of our medical resources? Many IVF clinics do not see age as a limiting factor, even though it is very difficult for women over the age of 45 to become pregnant in this way. Indeed, although some clinics argue that it is immoral to have a baby knowing that you will probably die before the child reaches adulthood, other clinics do not discriminate against any paying customer. Thus, in 2005 a 66-year-old Romanian woman gave birth to an IVF baby (Bjerklie 2005).

Does the state or a country have the right to limit assisted reproductive technology to married couples? To wealthy couples? To heterosexual couples? To women who are younger than 40? To people who have excellent health insurance? Right now, it is strictly a market economy, and there is very little regulation put on any physician who wants to start a clinic. If a couple knows it might be possible to have a genetically related child *if only they were wealthy enough*, does this frustrate more people than it helps? If curing infertility rather than simply making money is the goal, then should the focus be on a high-tech ART medicine, or should more be spent on public health efforts to eliminate sexually transmitted diseases—one of the leading causes of infertility? There are certainly numerous ethical issues raised by the economics of the new reproductive technologies.

Is there a "right" to have a genetically related child?

Is the desire to have a biologically related as opposed to an adopted child a definitive human biological urge, or is it being manufactured (or at least intensified) by advertising done by fertility clinics competing with each other in the present market?

There is irony in the fact that, at the same time a woman's right *not* to have an unwanted child is paramount in the abortion issue, so many couples' lives are being overwhelmed by their desire for a family they can't have naturally. And in our overpopulated world, millions of children are in desperate need of loving homes. Adoption seems to be a perfect solution to this social dilemma. But many couples who seek IVF are not inclined to adopt a child, even if the procedure fails.

> It's easy to love a baby, any baby, and yet so many of us yearn to love babies that we can claim credit for, that are ours, that we have made.
>
> NATALIE ANGIER (1999)

There are evolutionary premises to support the argument that a person's perceived need to propagate his or her own genes is not manufactured. Studies of many animal species, from social insects such as ants (an extreme example) to many bird and mammal species, show that genetic kinship is a driving force in animal behavior. E. O. Wilson's classic work *Sociobiology* (1975) documents this fact with meticulous scholarship. And Richard Dawkins' *The Selfish Gene* (1976) puts forth the highly controversial argument that evolution is driven not by an individual organism's need to survive, but by the need of that organism's *genes* to survive and be passed on to the next generation.

The arguments and raised hackles stemming from these books have occasionally reached the level of scientific "fistfights." But however we feel about the question of whether the behavior of other species has any bearing on the behavior of *Homo sapiens*, there is no doubt that most people feel a desire, perhaps even a drive, to reproduce. And it has been proposed that the pervasive belief in the supremacy of the gene that has arisen over the past 50 years has made some people feel it even more important to have a child that is "built from the beginning" by its parents (see Angier 1999).

Acknowledging these human needs, it is still a truism that "you can't always get what you want," and some people will inevitably have to accept that this truism applies to having a child of their own.

What is the legal status of an IVF embryo?

Is it an abortion to throw away the extra embryos that result from an IVF procedure? Many of these "excess" embryos are cryogenically frozen for potential future implantation (see page 68). Who has rights to keep these frozen embryos if the couple should divorce or die? And does the biological father have to make child support payments if the embryo is implanted and comes to term after a divorce? These are just some of the many legal questions that frozen embryos pose.

In U.S. courts, frozen embryos usually are considered property. This makes them contested items if the couple divorces. In February 2005, a couple undergoing IVF in Chicago found that one of their preserved fertilized eggs had been thrown out by mistake. The couple filed a wrongful death lawsuit, and won. The court ruled that this embryo was a human being even though it had not even been implanted into the uterus. Although this suit will probably be overturned, it highlights the disputed nature of frozen embryos.

Currently there are more than 100,000 frozen embryos in storage, and this number is growing at around 20,000 embryos per year (Andrews 1999). In Great Britain, the British Human Fertilisation and Embryology Act mandates that unused embryos be destroyed after a certain number of years.

In Conclusion

This chapter discusses only some of the questions posed by in vitro fertilization and related assisted reproductive therapies. Truth in advertising, the use of surrogate mothers, and the legal custody of babies created using sperm or egg donors are all issues that have taken up volumes of book chapters. Also, what are the ethics of selling sperm, eggs, embryos, or wombs? Who sets the prices, and under what condition can they be sold or "rented"?

Debora Spar of the Harvard Business School sees three options for American policy makers. The first is to establish an agency like Britain's. This, she thinks, is politically unfeasible because "in the current political climate, any federal foray into this area would probably fall prey to the politics of abortion, squeezing science in the process and limiting options for fertility treatment" (Spar 2005).

The second option is simply to keep the laissez-faire, market-based policies that we have now. This would mean that many infertile couples would not be able to afford such treatments, and would limit any debate about reproductive technologies such as genetic alteration and sex selection, which will be discussed in subsequent chapters. These issues strike at the core of what we mean by "motherhood," "fatherhood," "family," and even "baby."

The third option is to have minimal federal requirements to mandate what procedures can and cannot be performed (for instance, Congress may vote to ban reproductive cloning or the production of human-animal chimeras), while allowing individual states the power to ensure the safety and efficacy of each clinic and to establish which procedures should be covered by health insurance.

As we will see in the rest of this book, traditional ART is only the "tip of the iceberg." Prenatal sex selection, cloning, stem cell propagation, and genetic engineering are all possible, and society will have to decide who (if anyone) gets to use these techniques.

UNIT 3

Should We Select the Sex of Our Children?

The final aim of all love intrigues, be they comic or tragic, is really of more importance than all other ends in human life. What it turns upon is nothing less than the composition of the next generation.

A. SCHOPENHAUER
(QUOTED BY CHARLES DARWIN, 1871)

CHAPTER 5

The Genetics of Sex Determination

Chapter 3 described the details of mammalian fertilization, the process that transmits genes from parent to child and which activates development in the newly fertilized egg. Fertilization is also the time when the sex of a child is to a large degree determined.

We learned in Chapter 3 that fertilization initiates the formation of a new organism with a unique genome of 46 chromosomes—two copies each (one from the male parent and one from the female) of 23 chromosomes. In all but one instance, the chromosomes are for the most part equal: the two copies of chromosomes 1 through 22 will contain the same genes no matter which parent provided them (although the *information* in the gene may be different; the genes that determine eye color, for example, may specify brown or blue). These 22 numbered chromosomes are the **autosomes.**

The "twenty-third chromosome," however, is a different story—or, more accurately, two different stories. These **sex chromosomes** are not referred to by number but are designated X and Y. The genes carried on the X and Y chromosomes are very different, and some of them specify whether an embryo develops to be a male or a female.

Primary Sex Determination

Primary sex determination is the determination of whether an individual's **gonads** (sex organs) become testes or ovaries. In humans, primary sex determination is genetic, based upon which sex chromosomes are present. In females, the two sex chromosomes are both X chromosomes (XX). Males have one X chromosome and one Y chromosome (XY).

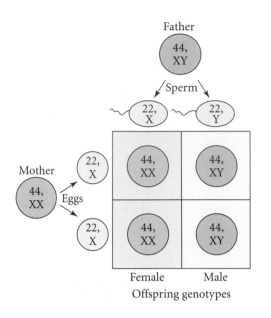

Father

44,
XY

Sperm

22,
X

22,
Y

Mother

22,
X

44,
XX

Eggs

22,
X

44,
XX

44,
XY

44,
XX

44,
XY

Female   Male

Offspring genotypes

FIGURE 5.1   Chromosomal sex determination in humans. Females produce eggs containing 22 autosomes and one X chromosome. Males produce two types of sperm, half of which contain 22 autosomes and an X chromosome, with the other half containing 22 autosomes and a Y chromosome. Sex determination depends upon which sperm meets the egg. The resulting zygotes will have 44 autosomes and be either XX (female) or XY (male). The total number of chromosomes, regardless of sex, will be 46.

Each of the haploid gametes formed via meiosis (see Figure 3.1) has only one sex chromosome. In eggs (the gametes formed by a female), the single sex chromosome will always be an X. When sperm (the male gametes) are formed, half the sperm will have an X chromosome and half will have a Y chromosome. When the sperm and egg unite, the conceptus will be female if the sperm that fertilizes the egg contains an X chromosome; the conceptus will be male if the sperm contains a Y chromosome (Figure 5.1). Thus, every human cell has at least one X chromosome, and the X chromosome is essential for cell function in both males and females. The Y chromosome, however, is strictly a sex chromosome: it carries genes that are needed for testis and sperm formation.

Formation of the gonads

The development of the gonads is unlike the development of any other organ. All other embryonic organ rudiments can differentiate into only one type of organ: a lung rudiment can become only a lung, and a liver rudiment can develop only into a liver. The gonadal rudiment, however, can develop into either an ovary or a testis. The path of differentiation taken by this rudiment determines the future sexual development of the organism. But before embarking on its sexual development, the human gonads first develop through a **bipotential stage**, during which time they have neither female nor male characteristics. The bipotential stage appears during week 4 of development and lasts until about week 7. At this stage, the gonadal rudiments are two pairs of **genital ridges**, one on each side of the lower abdomen. Each genital ridge is made up of an internal compartment of loose cells and an external compartment of tightly connected cells called the **sex cords**.

The formation of ovaries and testes are both active, gene-directed processes. At week 7, the primary sexual decision is made (Figure 5.2). *If the gonadal cells have the genotype XY*, testes are formed. The sex cords (the cells that will form the tissues that hold and nurture the germ cells) grow

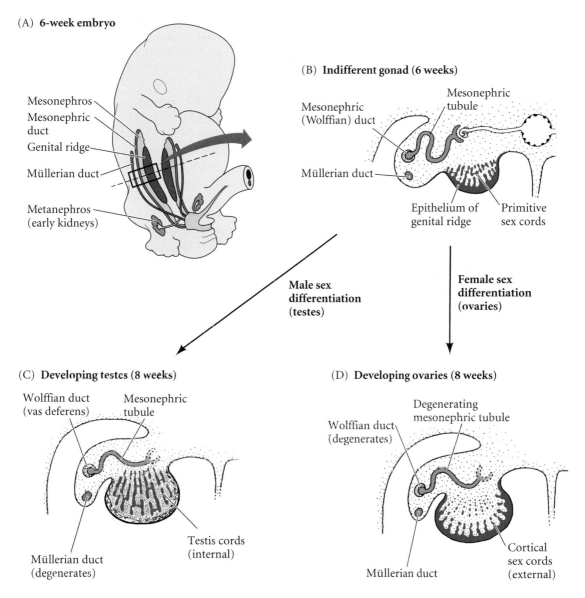

FIGURE 5.2 Differentiation of human gonads. (A) Diagram of a 6-week embryo, showing the position of the undifferentiated gonads. The mesonephros is an embryonic excretory organ. (B) Transverse section through the genital ridge of the 6-week bipotential gonad. (C) In testis development, the primitive sex cords develop inwardly, losing contact with the outside. They differentiate, connecting to the Wolffian duct such that the sperm can leave the body through the penis. (D) Ovary development in an 8-week human embryo. The internal primitive sex cords degenerate, but new sex cords take their place on the outside of the genital ridge.

inwardly. Gamete precursor cells migrate into these tubules and will stay dormant until puberty. The internal male sex cords become the **Sertoli cells** of the seminiferous tubules and the loose tissue around them becomes the **Leydig cells**. At puberty the tubules will open up and the gamete precursor cells will differentiate into sperm.

If the gonadal cells have the genotype XX, the bipotential gonads become ovaries. The sex cords develop in the periphery of the gonad, and the germ cells migrating into them will become eggs. The surrounding cortical sex cords will differentiate into **granulosa cells**. The loose cells of the ovary differentiate into **thecal cells**. Together, the thecal and granulosa cells will form the **follicles** that envelop developing eggs. Each follicle will contain a single egg precursor cell. Thus, in females, the gametes (oocytes) will reside near the *outer* surface of the gonad (where they can be ovulated into the oviduct), while in males the gametes (sperm precursors) reside in the *internal* part of the gonad (and need sperm ducts to get them to the outside).

Sex-determining genes

The "testis-forming" gene of the Y chromosome has been identified as the *SRY* gene (*SRY* stands for "sex-related gene of the Y chromosome"). XY individuals with mutations in the *SRY* gene develop ovaries rather than testes. In females, the *DAX1* gene on the X chromosome and the *WNT4* gene appear to be important in forming the ovaries and preventing testes from forming.

In females, two X chromosomes (in the absence of a Y chromosome) are necessary for the production of ovaries. Individuals with only one X chromosome and no Y chromosome (designated "XO") develop into sterile females whose ovaries fail to form properly.

The male and female germ cells

We mentioned in Chapter 1 that the **germ cells**—the precursors to the eggs and sperm—are set aside early in development. As the gonads are being formed, these gamete precursors migrate into the gonads. There are several differences between the male and female gametes as they reside in their respective gonads.

The female gamete precursor cells in the ovary initiate meiosis only once—while they are still in the embryo. They can stay in an immature state for decades, with one oocyte maturing to ovulate each month after menarche (the beginning of menstrual periods at puberty). Most of a woman's oocytes die, however, and relatively few of them are ovulated.

The male gamete precursor cells of the human testis, on the other hand, become **stem cells**. A stem cell is a cell that can divide so that one of the products of division is a more differentiated cell type (such as an immature sperm cell), while the other product is another stem cell just like the original cell. Thus, in human males, meiosis in the sperm begins at puberty, but new sperm are made continually throughout the man's life.

Secondary Sex Determination

Primary sex determination, then, is concerned with the sexual specification of the gonads (testes and ovaries) and germ cells (sperm and egg precursors). **Secondary sex determination** involves the sexual specification of the rest of the body. In humans, secondary sex determination is accomplished through hormones secreted by the gonads. Human secondary sexual determination specifies the form of the external genitalia; it also specifies that men and women have different distributions of body fat and body hair, as well as differences in muscle mass, voice tone, and pelvic bone structure.

Secondary sex determination also specifies either a male or female duct system. The early human embryo not only has a bipotential gonad, but also has two potential duct systems. It has a **Wolffian duct** that can become the male epididymis and vas deferens (the tubes that carry sperm to the outside of the body); and it has a **Müllerian duct** that can differentiate into the oviducts, uterus, cervix, and upper vagina. Figure 5.3 summarizes the different development of these two duct systems.

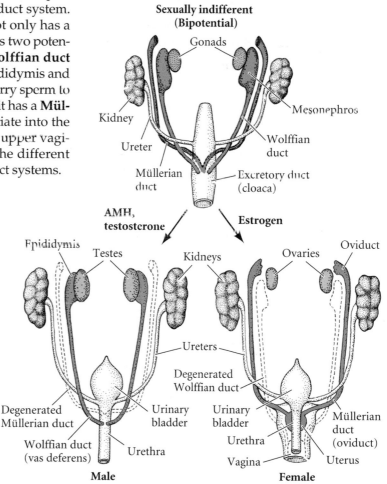

FIGURE 5.3 The development of the gonads and their ducts. Both the Wolffian and Müllerian ducts are present at the bipotential gonad stage. Without testosterone, the Wolffian duct and mesonephros degenerate, and estrogen causes the Müllerian duct (orange) to differentiate into the female genitalia (right). The testes make two major hormones. The first, anti-Müllerian duct factor (AMH), causes the Müllerian duct to regress. The second, testosterone, causes the differentiation of the Wolffian duct (blue) into the male internal genitalia (left).

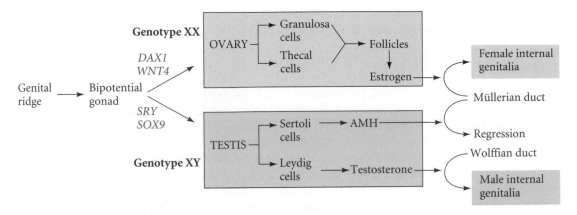

FIGURE 5.4   A simplified view of the cascade of events believed to lead to the formation of the male and female phenotypes in human sex organs. The bipotential gonad appears to be moved into the female pathway (ovary development) by the *WNT4* and *DAX1* genes. The ovaries make thecal cells and granulosa cells, which together are capable of synthesizing estrogen. Estrogen stimulates the development of the Müllerian ducts into the female genitalia. In the male, the *SRY* gene (located on the Y chromosome) activates the *SOX9* gene, which organizes the gonads to become testes. The Sertoli cells of the testes produce anti-Müllerian duct hormone (AMH), while the Leydig cells produce testosterone that stimulates the Wolffian duct into the male pattern.

If the embryo is XY, testes form and secrete two major hormones. The Sertoli cells secrete **anti-Müllerian duct hormone (AMH)**. AMH, as its name implies, destroys the Müllerian duct (which would otherwise become the female reproductive tract). The Leydig cells secrete a second hormone, **testosterone**. Testosterone promotes the differentiation of the Wolffian duct and its associated organs, such as the prostate gland. Testosterone also inhibits the development of breasts, stimulates the formation of the penis, and allows the testes to descend into the scrotum.

If the embryo is XX, the gonadal ridges develop into ovaries. The ovaries produce **estrogen**, a hormone that enables the Müllerian duct to develop into the uterus, oviducts, and upper end of the vagina. In the absence of adequate testosterone, the Wolffian duct of females degenerates. Estrogen also stimulates the onset of menstruation at puberty and, along with other hormones, regulates the cycle of ovulation. Figure 5.4 summarizes the interactions that result in physical and anatomical differentiation of the sex organs.

Sex-Linked Diseases

Every female cell has two X chromosomes, whereas a male cell has only one. Moreover, the X chromosome, unlike the Y, contains genes that are used in all the cells of the body. Therefore, if there are mutant genes on the

X chromosome, men are much more likely to be affected than women. For instance, Duchenne muscular dystrophy (DMD) is caused by the mutation of a gene that is normally active in muscle cells. Because this gene is on the X chromosome, it is referred to as an **X-linked gene**. A woman can have the mutant *DMD* gene on one of her X chromosomes and yet not have the disease, because she has a "good" (i.e., normally functioning) copy of that gene on her *second* X chromosome. However, although half of her eggs will transmit the normal *DMD* gene, the other half will carry the mutant gene (Figure 5.5).

If an egg with the mutant *DMD* gene is fertilized by a sperm with a normal X chromosome, the resulting XX girl will be normal (although she will still carry the mutant gene, as her mother did, and can pass it on to her offspring). However, if the egg with a mutant *DMD* gene is fertilized by a sperm carrying a Y chromosome, the resulting boy, with only the mutated X chromosome, will have no "good" *DMD* gene to balance the mutant gene and will have muscular dystrophy. Therefore, most deficiencies of X-linked genes are seen only in males.

Because females have two X chromosomes in each of their cells while males have only one, each female XX cell will *inactivate* one of its X chromosomes. Thus each cell expresses the genes on only one X chromosome. This **X-chromosome inactivation** is random: each cell has an equal probability of inactivating its paternally derived or its maternally derived X chromosome. If a female is heterozygous for a mutant X-linked gene, her tissue will usually function normally, since the tissues (made up of many cells) can usually work fine with only half the "good" protein product (that is, the product made by the normal X-linked gene). So even though X-linked diseases are almost always expressed by males, a female can manifest the disease if for some reason too many of her "good" genes are inactivated (a rare situation), or if she inherited the mutant gene from both parents (which would mean her father expressed the condition and her mother was a carrier).

FIGURE 5.5   A female may carry a disease-causing mutant gene on her X chromosome and yet be unaffected, because her second X chromosome has the normal variant of this gene and can compensate for the mutant gene. However, at meiosis, she makes two kinds of eggs, one whose X chromosome carries the mutant gene, and the other kind carrying the normal gene. If a sperm with a Y chromosome fertilizes the egg containing the mutant X chromosome, the boy will be affected by the disease.

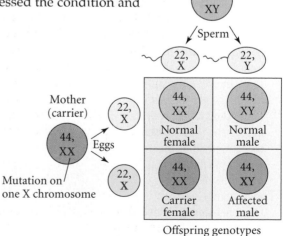

Androgen Insensitivity Syndrome

One of the most interesting sex-linked conditions is androgen insensitivity syndrome. There is a gene on the X chromosome that makes a testosterone receptor protein—a protein that binds to testosterone and carries it into the cell nucleus. Without this receptor protein, testosterone cannot function.

Imagine an embryo that is XY, but with a mutation on its X chromosome in the gene for the testosterone receptor. This embryo makes testes (because it has a functional Y chromosome with a normal *SRY* gene); the testes in turn produce their two hormones, AMH and testosterone (see Figure 5.4). The anti-Mullerian duct hormone from the Sertoli cells functions normally: the Müllerian ducts disintegrate and the fetus has no cervix, uterus, or oviduct. But because of the mutation in the testosterone receptor protein, the testosterone made by the Leydig cells cannot function.

Now enter the influence of estrogen and related hormones. Estrogens are found in both male and female embryos; they are made by the adrenal gland, the testes, and the mother. In androgen-insensitive embryos, under the influence of estrogens the breast buds will be programmed to develop in the female manner, as will the hair and body fat distributions. Indeed, the testes (which are formed inside the abdomen and whose descent is regulated by testosterone) will not become external. When

such a baby is born (and until she fails to have menstrual periods as a teenager), the individual looks and behaves like a normal woman; however, her anatomy prevents her from having children.

Persons with androgen insensitivity syndrome sometimes undergo surgery to extend the vagina and remove the nonfunctioning internal testes (which are prone to become cancerous if not removed).

If one views being "female" as a *social* category, then persons with androgen insensitivity syndrome are normal females, even though they carry a Y chromosome in every cell. If "female" is viewed as a strictly *genetic* category, however, these same people are abnormal males whose development has been arrested by a genetic mutation. If femaleness is viewed as a *functional* category, then these same people are abnormal females who lack a uterus and cannot bear children. Androgen insensitivity syndrome is thus one of several **intersex** conditions that make us ponder what we think of as "normal." How biology and society interact in determining what is "normal" is discussed in Chapter 13.

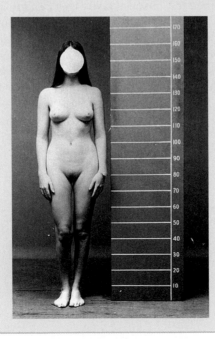

ANDROGEN INSENSITIVITY SYNDROME. Despite their XY genotype and the presence of internal testes, people with this X-linked syndrome develop the secondary sex characteristics of a normal female. (Photograph courtesy of C. B. Hammond.)

Prenatal Diagnosis and Preimplantation Genetics

One of the consequences of in vitro fertilization (see Chapter 3) has been our ability to detect genetic mutations early in the embryo's development. Many genetic diseases can be diagnosed before the baby is born; such **prenatal diagnosis** can be done by chorion biopsy at 8–9 weeks of gestation or by amniocentesis at around the fourth or fifth month of pregnancy. **Chorion biopsy,** or **chorionic villus sampling,** involves taking a sample of the placenta, while **amniocentesis** involves taking a sample of amnionic fluid. In both cases, fetal cells can be analyzed for the presence or absence of certain chromosomes, genes, or enzymes.

However useful the procedures of prenatal diagnosis are in detecting genetic disease, they carry with them a significant concern: If the fetus is found to have the genetic disease, the only means of preventing it is to abort the pregnancy. The waiting time between the knowledge of being pregnant and the results from the amniocentesis or chorion biopsy can create a "tentative pregnancy," a stressful period during which many couples do not announce their pregnancy for fear that it might have to be terminated (Rothman 1994).

One way around this difficult choice would be to screen the embryonic cells before the embryo is even in the womb. Such screening is indeed possible during the period between in vitro fertilization and the implantation of the embryo into the uterus. This ability to test an in vitro embryo is the basis of a new area of medicine called **preimplantation genetic diagnosis,** or **PGD**. PGD seeks to test for genetic disease prior to the embryo's entering the uterus.

Preimplantation screening procedures are performed on embryos created by in vitro fertilization while the embryos are still in the petri dish. At the 4- to 8-cell stage of the embryo, a small tunnel is made in the zona pellucida, and a micropipette removes two blastomeres (Figure 5.6). Since the

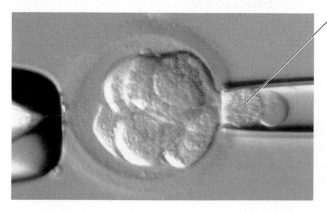

Blastomere removed for genetic analysis

FIGURE 5.6   Preimplantation genetic diagnosis (PGD). This procedure is performed on one or two cells taken from an early (4- to 8-cell stage) embryo. Microscopy can tell whether the chromosomes are normal, and other lab tests reveal whether certain genes are present, absent, or mutant. (Courtesy of the Institute for Reproductive Medicine, Livingston, NJ.)

mammalian embryo can easily replace these cells, their removal does not endanger the embryo. (This plasticity of early embryonic blastomeres was mentioned in Chapter 1 and will be discussed more fully in Chapter 7). The isolated cells can be tested immediately for the genes of interest, with results that are often available within two days. The presumptive "normal" embryos are implanted into the uterus, while presumptive defective embryos are discarded. For many couples, it is easier to consider implanting only those embryos that most likely will be healthy than aborting a fetus, even if the fetus is in all probability seriously impaired.

Sex Selection

Besides screening for genetic defects, preimplantation genetic diagnosis allows couples to know another piece of information that used to have to await the baby's birth: these tests can reveal the embryo's sex. Some parents want to know this information; some do not. Such knowledge raises the issue of parents choosing the sex of their offspring in advance.

There are probably very few people who would condone aborting a pregnancy simply because the embryo was the "wrong" sex (although this is known to happen, as discussed in Chapter 6). However, using PGD in conjunction with in vitro fertilization can allow a couple to choose their child's sex by having only embryos of the desired sex implanted. Different countries and even different hospitals have varying regulations as to whether they permit preimplantation diagnosis solely for the purpose of sex determination.

Another way to accomplish sex selection is through **sperm sorting**. The basis for this procedure, which uses a biomedical technique called flow cytometry, is that the X chromosome is substantially larger than the Y chromosome. Unlike preimplantation diagnosis, sperm sorting is *preconceptual*: sperm selection occurs prior to fertilization and thus this method does not involve either the abortion or destruction of "wrong-sex" embryos. It is an expensive procedure, however, and is not likely to be covered by insurance; thus it is reserved for those wealthy enough to afford it.

Sperm sorting using flow cytometry was developed by a private laboratory in 1989. The technique exposes a sperm sample to a DNA-binding fluorescent dye. The female-producing sperm contains more DNA than male-producing sperm (because the X chromosome is much larger than the Y chromosome), and thus absorb more dye. The dye-injected sperm sample is then passed through the flow cytometer, a laser-based device that activates the dye and then separates sperm one at a time according to brightness density, separating the brighter (X-chromosome) sperm from the less fluorescent (Y-chromosome) sperm (Weaver 1999).

Sperm separation is not 100 percent effective, but it does result in a sperm sample that is significantly enriched for either X-bearing or Y-bearing sperm. The enriched sample of choice can then be used for artificial insemination or in vitro fertilization, allowing a couple a significantly increased chance that the embryo produced will be of the desired sex. If the desired result is a girl, there is about a 90 percent success rate. Selection for a boy is less reliable, with about a 75 percent success rate (Ramachandran 1999). This disparity occurs because an X-carrying chromosome that appears even slightly dimmer than the norm will be sorted as an male-producing sperm.

Sex selection, almost always using preimplantation genetic diagnosis, is widely used therapeutically as a way of preventing X-linked diseases. It is a boon to women who know, or who have reason to believe (from family medical history), that they carry an X-linked mutation. However, the wider potential for sex selection is in fact its use as a method of family balancing: if a couple can have children of the desired sex in the sequence they prefer, proponents argue, it will result in smaller families (a good thing in an over-populated world) and thus less economic and emotional burden for many. Opponents of sex selection, however, point to its probable use to prevent the birth of girls in cultures where women are regarded as less important than men. This concern and other issues are the subject of the next chapter.

CHAPTER 6

Arguments For and
Against Sex Selection

D iscussions of cloning and stem cell research involve setting limits on technologies that do not yet exist. However, the debate on whether sex selection is an ethical practice concerns technology that is already perfected. It is possible, through preimplantation genetic diagnosis (PGD), to determine which 4- or 8-cell human embryos are male and which are female, and to implant into the uterus only embryos of the desired sex.

Preimplantation diagnosis is widely used to prevent the birth of sons in women known to be carriers for X-linked lethal or debilitating diseases (see pages 88–89). For instance, a woman who is a carrier for the X-linked conditions hemophilia or muscular dystrophy may not want to suffer through the "tentative pregnancy" associated with amniocentesis or chorionic villus sampling, and then have to decide whether or not to abort the male fetus should it prove to have the disease. Such women can instead undergo in vitro fertilization and have only XX embryos implanted; the female infants should be unaffected by X-linked diseases (although they have a 50 percent chance of being carriers).

Using PGD for the diagnosis and prevention of serious medical conditions seems appropriate to most people (especially if they are not morally opposed to IVF in general; see Chapter 4). However, if a couple have a son and now want a daughter, should they be allowed to use the new sex selection technologies to assure the outcome? If a couple has three daughters and the husband desparately wants a son to "carry on his name," should this family be allowed access to the technology? Empires have fallen from the lack of a male heir; even today there is debate in Japan about how far the royal family should go in trying to produce a male to sit on the Chrysanthemum Throne, since the Crown Prince and Princess have only daughters. Moreover, there are some societies where the cultural pressures favoring males are so strong they can drive parents to abandon or kill their female offspring. What would be the result of readily available, reliable sex-selection technology in such countries?

A Brief History of Sex Selection Practices

Throughout history, couples have made attempts to control the sex of their offspring. In almost all cases, males have been preferred due to social circumstances favoring the male's greater ability to earn money for the family, provide manpower to armies, work the fields, care for aging parents, inherit possessions, and carry on the family name.

There seems no end to the techniques that parents have tried in order to influence the sex of their child. Aristotle, who believed that high temperatures produced male children, counseled men (especially elderly men) to have intercourse in the summertime if they wished male heirs. Ligating testicles (the right one was believed to contain the male-producing semen), eating certain foods, mating at particular times in the woman's menstrual cycle, and wearing boots to bed have all been promoted as ways to influence the sex of one's offspring. Such folkloric solutions can still be found on today's Internet, and these are every bit as reliable as the ancient prescriptions—that is to say, they produce the desired result about 50 percent of the time.

As twentieth-century scientists learned more about the intricacies of reproduction, scientifically founded but still marginal techniques were proposed to increase the odds of conceiving one sex or the other. In 1970, David Rorvik and Landrum Shettles proposed a technique based on the swimming speeds of sperm. Because male-producing sperm carry a Y chromosome, which is smaller than the X chromosome carried by female-producing sperm, Rorvik and Shettles reasoned that the Y-bearing sperm should be able to swim faster along the female reproductive tract. Therefore they recommended that to have intercourse close to the time of ovulation would give the faster, Y-bearing sperm the advantage and increase the odds of conceiving a son. Conversely, their theory holds that having intercourse several days before ovulation tips the scales in favor of conceiving a girl. A second hypothesis from Rorvik and Shettles claimed that using a weakly acidic douche to wash out the vagina and cervix just before having intercourse would increase the chance of having a girl, whereas using an alkaline douche would increase the chance of a boy (Rorvik and Shettles 1970).These maneuvers give, at best, a minute statistical advantage to one type of sperm or the other; couples using them often end up with a child of the unintended sex.

When recipes for sex determination fail to produce the desired results, couples throughout history have turned to more direct methods of sex selection—namely, infanticide and abortion. There is evidence that infanticide was practiced in ancient Greece, throughout the early Roman Empire, and in the Arab world to select for male children (Warren 1985). And there is evidence that both infanticide and sex-specific abortion occur in modern countries such as India and China, where extreme cultural pressure to give birth to males collides with increased governmental and economic pressure to have small families; this dilemma will be discussed at length in the next section.

In the 1970s, Gametrics, a Montana-based lab under the aegis of Roland Ericsson, developed a method of sperm separation for use in livestock breeding. (In dairy farming, for example, male cattle are superfluous and expensive, since nearly all commercial cattle breeding is through artificial insemination.) Like that of Rorvik and Shettles, Ericsson's method relied on sperm swimming speed: sperm containing Y chromosomes presumably swim slightly faster than those bearing X chromosomes.

Gametrics separated the X- from the Y-bearing sperm by passing the sample through viscous (sticky) materials such as albumin, Percoll, or complex carbohydrates to create a "chemical obstacle course" that exaggerated the difference in their swim speeds ("Unnatural Selection," 1993). This method of sperm separation, the only one available until a few years ago, was only capable of "enriching" semen, or offsetting the ratio of X-bearing to Y-bearing sperm, by about 10 percent.

In the early 1990s biotechnology laboratories in the United States developed the flow cytometry technique of sperm sorting (see pages 92–93). Sperm sorting, for those who can afford it, can now be quite effective. And using the preimplantation genetic diagnostic procedures described in Chapter 5 allows one to choose an embryo's sex with virtually 100 percent certainty. Sex selection is now possible, but to what extent is it a good idea? There are a number of serious issues to be considered.

Pressures for Sex Selection

Many factors play into a parent's desire for children of one sex over the other, and these factors vary in different parts of the world. In most of the world—and particularly the Far East—cultural, personal, and economic issues merge to drive the balance in favor of male babies (Robertson 2001). Much of this attitude is deeply ingrained in the cultures that share it, and many regions have long histories of preferring males. On the other hand, many Westerners who promote sex selection see it as a way to achieve what is known as family balancing, since in these countries it appears the preference in most families is to have children of both sexes (Kalb 2004).

Economic and cultural pressures and the gender gap

Matters of finance and family economics are perhaps the most overt forces working around the globe to drive sex selection. In India, for example, daughters are viewed as an expense, while sons are seen as a financial asset (Ramachandran 1999). Daughters need to be provided with dowries when they marry,* and patrilineal tradition ensures that many women join their husbands' families upon marriage and thus are no longer available to care for their own parents. Indian culture is also influenced by Sanskrit literature, which is interpreted as saying that the main purpose of marriage is to give birth to a son (Mudar 2002). In some parts of India, particularly poor rural areas, it is not uncommon for the midwife, or *dai*, to hold a female newborn upside down by the waist, give it a jerk to snap the spinal column, and pronounce a stillbirth (Carmichael 2004).

Among the Chinese, the kinship system emphasizes paternal descent. Patrilocal residence is the norm, and Chinese parents rely on their sons for support in their old age. Because the perceived need for a son is so high, and because the Chinese government places heavy financial burdens on families with more than one child, infanticide of females is believed to be relatively common. Female infants are often left in the streets or on doorsteps of orphanages (Li 1991; Vines 1993; Winkvist and Akhtar 2000).

In the Malay culture of Southeast Asia, on the other hand, matrilineal kinship is practiced in some areas, while bilateral kinship appears in others. In these regions there is no consequence of surnames for lineage, and girls are thought to take better care of their parents, so if any sex preference arises, it is slightly in favor of females (Pong 1994). The Second Malaysian Family Life Survey in 1988 examined Chinese, Indian, and Malaysian populations living together in Malaysia. The survey found that Indians had the strongest preference for boys, followed by the Chinese with a moderate preference, and the Malays, who had no preference for one sex over the other (Pong 1994).

The overwhelming preference for male children among parents in India and China has had sobering results. The technologies of amniocentesis and ultrasound, vital medical technologies for the health of many women and infants, both allow sex identification within the first trimester of pregnancy—well within the legal limits of abortion in most countries. If a woman requests an abortion, it is impossible in most cases to differentiate whether she does not want a baby at all, or does not want a baby of that particular sex.

In wealthy areas of India, where ultrasound and amniocentesis are available, abortion rates of female fetuses is disproportionately high, even though sex selection by any means is illegal and the Hindu religion predominant in the area strictly proscribes abortion (see Chapter 2). In Bom-

---

*The government of India, recognizing the severity of the sexism the dowry system has engendered, has begun instigating policies to create a movement away from dowries and toward "bride money." Bride money is paid by the groom to the bride's family, and by encouraging this tradition, the Indian government hopes to combat the idea that daughters are a financial liability (President's Council on Bioethics 2003).

bay, a 1985 survey found that 96 percent of aborted female fetuses were aborted after amniocentesis revealed their sex (Ramachandran 1999). One study found that out of 8,000 reported abortions, 7,999 of them were of female fetuses (Roberts 2002). In 1994, the Indian government passed the Pre-Natal Diagnostic Techniques Act, which attempted to regulate prenatal testing such as sonography and made it illegal to use such procedures for sex selection. However, it seems that, in spite of the act, sex-selective abortion following sonography remains widespread (Shete 2005).

Because of the overwhelming preference for boys in most Eastern cultures, the **gender gap**—the extent to which the sex ratio of males to females deviates from the theoretical norm of 100:100—has become a major issue in many countries (Macklin, 1995; Satpathy and Mishra 2000). The 2001 Indian census showed that the country's sex ratio rose to 108:100 during the 1990s (Mudar 2002). In China, as of 1996, males outnumbered females by 36 million (Cardarelli 1996).

In some regions (most often rural areas), gender gaps have led to a generation of young, single males who have no prospects of marrying. This demographic group has been around long enough in some countries to earn its own term. In China, the young men are known as *guang gun-er* ("bare branch") because they represent "branches of a family tree that would never bear fruit because no marriage partner might be found for them" (Hudson and Den Boer 2004). Studies have shown that the *guang gun-er* commit a disproportionately high fraction of the crime in their respective areas.

The cultural situation in countries like India and China presents a real dilemma for proponents of sex selection technologies. Even if it were economically feasible to make the current (expensive) technologies widely available there, and even if large numbers of people in these cultures could be brought to accept the loss of privacy and invasiveness of these techniques, would the results really be beneficial? Parental choice would certainly help curb infanticide, abandonment, and the huge number of sex-based abortions. But would the technology allow the gender gap in these countries to skew even more strongly and dangerously to an overabundance of males?

Family balancing

Gender gaps in the Western world are not nearly as pronounced as in Asia. Indeed, a 2001 article in the British Medical Journal claimed that "there are studies [in England] which show universally that there is no preponderance of one gender" and that "in Western society there are as many couples who want the girl as there are who want the boy" (Gottlieb 2001). Similarly, a 1993 study by the American Association for the Advancement of Science and a 1999 report in Canada found that just as many women preferred girls as boys, and that most had no preference one way or the other (Vines 1993; McDougall et al. 1999).

Although different studies reveal slightly different statistics, it seems likely that in many Western countries, the preferences for girl-versus-boy would balance out over the long run. The goal of sex selection in these countries would be **family balancing**—the ability to choose to have a family of a desired composition (often perceived to be two children, one of each sex, with a slight preference given to the boy being born first) (Silver 1998; McDougall 1999).

Where biases have been found, they are often tightly linked to the number and sex of children already born to the family. For example, an extensive study done in the U.S. between 1970 and 1975 showed the strongest sex preferences to be in women who already had two children of the same sex (Figure 6.1; Pebley and Westoff 1982). A 2005 study appears to confirm that, in the United States at least, the majority preference is not for one sex over the other, but for a family with children of both sexes (Jain et al. 2005). And the 1999 Canadian study showed Canadian women to prefer sons as firstborn children; however, when Canadian couples were interviewed as a unit, the preference disappeared, indicating that studies targeting only women may not accurately represent gender preferences of entire populations (McDougall et al. 1999).

With the widespread use of sex-selection technology, a slight gender gap might initially be created, but some predict that over time the gap would close (Smith 1993). Rather than an overall change in numbers of males and

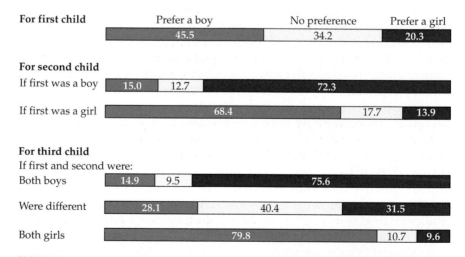

FIGURE 6.1   Preferences of married women in the United States for the sex of their next child. The women were surveyed over a number of years (1970–1975). The numbers are the percentages of women preferring a boy (blue bars) or a girl (red bars), and those expressing no preference one way or the other (yellow bars). (Data from Pebley and Westoff 1982.)

females, some predict that "there would be a significant increase in the probability of the firstborn being male and the second child being
female, and a large drop in the probability of
both being the same sex" (Westoff and Rindfuss
1974). Even this, however, could have a major
impact on a society's structure.

> [T]he scarcer sex would gradually become the more valued one, and the sex ratio would tend back to unity.
>
> BRITISH MEDICAL JOURNAL (1993)

Ethical Views on Sex Selection

Aside from serious questions about the impact widespread practice of sex
selection might have on the gender composition of different societies, many
people question whether or not it is really a good idea for people to exercise
this kind of control over choices that have heretofore been out of human
hands. Some think the technology would be beneficial, insuring smaller families in an overpopulated world and less tension in certain family situations.
Others think see the practice as sexist, antisocial, or otherwise immoral.

Attitudes toward sex selection vary among and within countries. In the
United Kingdom a recent survey found that 69 percent of interviewees
thought sex selection should be restricted in some way, and 80 percent
thought that sex selection for nonmedical reasons should not be permitted
at all (Ethics Committee 2001). Similarly, the authors of a 2003 German survey reported in *Human Reproduction* concluded that, even if they did not
condemn the use of prenatal sex selection by others, most Germans (92 percent of those surveyed) would not use it themselves (Dahl et al. 2003). In the
same survey, 58 percent of the respondents stated that they didn't care
which sex their offspring were, while 30 percent preferred a family with an
equal number of boys and girls. About equal numbers wanted a preponder-
ence of one sex (4 percent wanted more boys, 3 percent more girls) and 2
percent preferred only girls or only boys (1 percent each).

The idea of sex selection seems slightly more popular in the United
States. A 2005 survey of women being treated for infertility found that 41
percent of them said would make use of preimplantation sex selection if it
were offered to them (Jain et al. 2005). However, many people still disapprove of the ability to choose the sex of one's child. Indeed, in response to
the *Human Reproduction* survey, Brent Waters, a Christian theologian, wrote
that "most respondents have no intention of using sex selection techniques,
reflecting, I believe, a moral intuition that there is something inherently
wrong with the process itself" (Waters 2003).

Religious viewpoints

The question of sex selection using preimplantation diagnosis is inextricably tied with religious beliefs on when life begins, as addressed in Chapter

2. If the religious view is that life begins at conception, then PGD, which involves the the manipulation and disposal of zygotes and early embryos, is usually proscribed.

There is no definitive viewpoint among the various Christian denominations about the morality of sex selection. The Roman Catholic doctrine of strict interdependence between the acts of marital union and procreation eliminates any potential for sex selection. The benefits of PGD in certain circumstances are certainly acknowledged and often supported among some protestant denominations, although the use of this technology for nonmedical purposes is generally eyed askance.

Another perspective found in Western religions is apparently rooted in opposition to pagan practices of sex selection through infanticide. Infanticide was practiced widely in the pre-Christian Roman Empire, but was outlawed as Christianity came into power (Warren 1985). Similarly, the emergence of Islam in the Arab world suppressed infanticide as a means of sex selection in the Middle East. Prior to this point, bearing a female child was considered shameful; however, Muhammad, the prophet of Islam, attempted to convince the Arab people that the killing of any soul created by God was a sin, and that the souls of women are as valuable as the souls of men (Giladi 1990).

In 1983, Islamic scholars convened a seminar to discuss human reproduction (Islamic Organization for Medical Sciences 1983). At the time there was agreement among the participants that sex selection on a national level was unlawful; however, there was a split among the scholars as to whether PGD practiced on a case-by-case basis should be legal. Some thought the wishes of individual couples should be met, while others feared the skewed sex ratio that might result from the practice. There is still no unanimous decision among Muslim religious leaders regarding sex selection technology, with some being of the opinion that family balancing offers a valid point to support the practice (Al-Serour 2000).

However, the political and social differences between men and women in much of the Arab world continue to result in boys being the more desired offspring. For example, a recent decision in Dubai concluded that married men (but not married women) could legally be cloned (Andrews 1999). And Arab physicians confronted with a baby having ambiguous genitalia (see Chapter 5) have refused to perform surgery that would render the child a girl rather than a boy (Fausto-Sterling 2000). Thus sex selection in this culture retains the potential for population imbalance.

Jewish tradition maintains a very strong relationship with the field of medicine, in which it is thought that "he who saves one human life, is as if he saved an entire world" (Shalev 2003). For this reason, genetic screening, genetic engineering, and other genetic manipulations are allowed and even encouraged by Jewish teachings if undertaken to prevent disease and save lives (Rosner 1998). In fact, a campaign to screen potential parents for Tay-Sachs (a recessive mutation that is lethal to the child when both parents transmit the trait) has led to a 90 percent reduction in the occurrence of Tay-Sachs among Jewish infants born in North America. The Tay-Sachs program has been cited as a model of how genetic screening can help curtail certain devastating medical conditions (Mange and Mange 1999).

When life and limb are not in jeopardy, however, genetic screening is less extolled. In orthodox Judaism, a man must father at least one boy and one girl before he has fulfilled his duty to "be fruitful and multiply," and some have asked whether this task might be facilitated by sex selection. When questioned about this, however, most rabbis deem sex selection for nonmedical purposes unacceptable (Wahrman 2002). In a September 2003 meeting, the Israeli Ministry of Health ruled sex selection for nonmedical purposes to be illegal (Shalev 2003).

Ethical positions among the medical community

The official position of both the American College of Obstetricians and Gynecologists and the International Federation of Gynecology and Obstetrics opposes sex selection except for medical reasons. The American Society of Reproductive Medicine, however, takes the stand that preconception sex selection to achieve family balance—that is, to provide a couple with a child of a different sex than their existing children—is ethically sound as long as the techniques used are safe and effective (Wahrman 2002; Jain et al. 2005).

Dr. Jeffery Steinberg, head of the Fertility Institutes in Los Angeles, argues that much of the current disapproval is merely a result of the newness of the technology, and that as it becomes more ubiquitous, attitudes will tend towards acceptance (Kalb 2004). This trend has proven true among the geneticist community in the United States. In the mid-1970s—long before PGD was a reality—only 1 percent of medical geneticists approved of prenatal diagnosis for sex selection and the abortion of fetuses of the "wrong" sex; by 1989, a *New York Times* poll reported that 20 percent approved of the practice (Leo 1989). A 1985 study showed that 62 percent of U.S. geneticists would either perform prenatal diagnosis for the purpose of sex selection themselves or refer the couple to someone else who would (Wertz 1989).

Thus there appears to be widespread acceptance of therapeutic sex selection, and many members of the medical community also endorse family balancing. However, doctors must face the question of whether a potential parent has reasons beyond family balance for wanting to choose their child's sex. The possibility of "ulterior motives" is of deep concern to medical ethicists.

Gender stereotyping, discrimination, and "commodification"

Many ethicists believe that there is something inherently wrong in the ability to choose the sex of one's child, and they have articulated their arguments against those who call for wider distribution of the technology.

Rebecca Dresser, professor at Washington University Schools of Law and Medicine and a member of the President's Council on Bioethics, argues that sex selection can cause irreparable damage to the mother-child relationship—a relationship, she states, that should be one of unconditional love, not based on the child's gender (Human Genetics Alert Campaign 2002).

Another major issue is the perception that choosing one sex is rejecting the other. What message does it send to a mother's sons if she is willing to spend thousands of dollars just to ensure that she does not have another boy? What message does a little girl get from parents who are searching for ways to make sure their next child is a son? While some people point out that a person can choose one sex over the other without necessarily thinking that either sex is "superior," it is clear that societal discrimination goes beyond personal preference to forces much larger and more unwieldy (Ethics Committee 2001).

To choose to have a child of a selected sex relegates both parent and child to predetermined roles in the relationship, and pressure is placed on the child to behave in a certain gender-specific manner (Ethics Committee 2001; President's Council on Bioethics 2003). Gender stereotypes pervade society, and represent another major argument of those who assert that sex selection is unethical. The Human Genetics Alert Campaign dubs prenatal sex selection as "the exercise of sexism at the most profound level, choosing who gets born, and which types of lives are acceptable." Many ethicists see sex selection as treating embryos as commodities, and believe that this "commodification" of children is unacceptable (Roberts 2002). Michael Sandel of Harvard University and the President's Council on Bioethics argues against what he terms "product selection." When we select a product at the grocery store, we do so because we expect it to have certain characteristics and be of a certain quality. Similarly, in choosing a child' sex, we are expecting the child to conform to our preconceived ideas of how the particular sex behaves (President's Council on Bioethics 2003). Any variation may lead to disappointment.

> In how many cases where parents are "desperate for a girl" will they be hoping for a loud tomboy that grows up to be an engineer?
>
> Human Genetics Alert Campaign (2002)

In addition to the psychological and ethical concerns inherent in the idea of child commodification, the very term "commodification" connotes a for-profit industry built around sex selection. A survey conducted in 2001 by *Fortune* magazine found that 25–35 percent of prospective parents might consider sex selection (Wadman 2001). Even if only 2 percent of this 25 percent were to actually use the technology, pre-selection could easily become a $200 million per year industry in the United States. Theologian Brent Waters bemoans the "growing perception of children as commodities satisfying the desires of their parents," writing that "sex selection technology is

Preimplantation Genetic Diagnosis: Toward *GATTACA*?

Today it is possible to select the sex of one's child through preimplantation genetic diagnosis. With PGD, it is a simple matter to check a 4-cell embryo for the presence of a Y chromosome, which indicates the child is male. Potential parents can then choose to implant only those embryos with a Y chromosome if they want a boy, or only those without a Y chromosome if they want a girl. The same preimplantation techniques are also used routinely to screen embryos for the presence of genes that are known to cause certain deadly or seriously debilitating medical conditions, thereby circumventing the birth of children who would suffer from these conditions.

But what about the genes that are responsible for, say, height, or eye color, or curly hair? As researchers locate and describe more and more human genes (see Chapter 9), what if we can pinpoint genes that bias a person's intelligence, or athletic prowess, or sexual preference? It is certainly possible that we will someday—perhaps even soon—be able to select for other physical and mental traits in much the same way we can currently select for sex.

The 1997 movie *GATTACA* depicts a future society in which any imperfection of the human body or mind is abhorred and discriminated against. After all, if genetic techniques allow people to insure their child receives only the "best" traits, can't it be seen as antisocial not to take advantage of these procedures? Is the movie sheer science fiction, or could such a social scenario actually come to exist?

GATTACA's premise is that if a trait is perceived as undesirable, and if it then proves possible to prevent or eliminate that trait, society will inevitably discriminate against people who possess the trait, relegating even normal human variation (poor eyesight, short stature, heavy body frame) to the status of "diseases" (see Paul 1995).

It has been argued that the mere decision to prevent the birth of a child with a given trait—whether the trait is the child's sex or height or hair color—is in itself a form of discrimination. Others see nothing wrong with biasing one's odds of having a "normal" or "superior" child (Dahl 2003; Roberts 2003). It is not hard to see that the ethical arguments over preimplantation genetic diagnosis for sex selection may be just the tip of a very large iceberg.

but one more tool for developing a market in desirable children" (Waters 2003). Indeed, there are already numerous websites devoted to "selling" the procedure to parents, and radio, magazine, and newspaper ads flashily offer the services of various prenatal diagnosis clinics (President's Council on Bioethics 2003).

The Ethics Committee of the American Society for Reproductive Medicine tries to combat the "free market" mentality by arguing that, when the health of the child is not a risk, sex selection is "inappropriate control over trivial characteristics" (Ethics Committee 2001). And if health is not a concern in the arena of social sex selection, it can be argued that the process then becomes a matter of enhancement, and the door to "eugenics driven by the free market" is "thrown wide open" (Human Genetics Alert Campaign 2002).

Government Policy: Rights versus Regulation

Over 40 European nations and several countries in Asia (including India) have banned all nonmedical sex selection (Mudar 2002; Human Genetics Alert Campaign 2002).* In the United States, however, there has not even been much debate on the issue up to this point. The novelty and financial inaccessibility of the technology have kept demand for regulation low, and there are currently no federal or state laws that deal expressly with sex selection.

One argument by those in favor of prenatal sex selection in the United States invokes the concept of human rights. According to the U.S. Constitution, citizens have the "right to be free from undue interference," and some feel that sex selection of children is "a logical extension of parents' rights to control the number, timing, spacing, and quality of their offspring"(Wertz 1989). Rachel Remaley cites the 1972 Supreme Court decision upholding the "right to privacy and freedom from intrusion affecting the decision to bear or beget a child."[†] Remaley argues that if a woman currently has the right to choose whether or not to have a child, she should also be allowed to choose which sex the child will be (Remaley 2000). Others cite the Universal Declaration of Human Rights (1948), which includes among the enumerated human rights "the right to marry and found a family" (this particular passage in the Declaration arose in response to Nazi laws prohibiting reproduction by disabled people).

Others, however, feel that sex selection is not among the inalienable rights alluded to in the Declaration of Independence, nor is it included among the "reproductive rights" upheld in Supreme Court decisions. In response to those who cite the Universal Declaration, the Human Genetics Alert Campaign points out that people are already legally disallowed from certain potential reproductive activities, for example marrying close family members (Human Genetics Alert Campaign 2002). They contend that reproduction with whomever we choose through any technological means available is not necessarily a human right, and sex selection should be evaluated as a separate case.

Because the fear of new technology often increases the desire for its regulation, it is expected that legislation governing prenatal genetic diagnosis and sex selection in the United States will not be long in coming (Roberts 2002), although some feel that the regulation of PGD is outside governmental territory and should be done on an independent basis.

---

*For a concise summary of current international legal regulations regarding sex selection technology, see http://www.bionetonline.org/English/Content/db_leg2.htm#sex.

[†]The 1972 Supreme Court decision in *Eisenstadt v. Baird* struck down laws prohibiting the distribution of contraceptives to unmarried persons. It was one of a series of decisions on reproductive freedom that paved the way for *Roe v. Wade* in 1973.

One group asking for regulation of this technology has been a coalition of women's rights and technology assessment organizations. In 2002, they published a letter to the American Society of Reproductive Medicine in response to the Society's endorsement of prenatal sex selection (Center for Genetics and Society 2002). The letter expressed deep concerns about the inherent potential for gender discrimination posed by the practice of sex selection: "While motivations for desiring a child of a particular sex may vary, there are no non-sexist reasons for pre-selecting sex except in cases of preventing serious sex-linked diseases. This is true even in the United States, where economic and social pressures to raise male children are minimized in comparison to other societies." However, there is ambivalence among feminists over the question of sex selection (Macklin 1995). On one hand, feminists traditionally have defended a woman's reproductive liberty in all cases; on the other, some feminists uphold the argument of the coalition's letter, that sex selection will encourage sex discrimination.

In anticipation of laws that might be enacted, Liu and Rose (1996) proposed these guidelines for family balancing clinics:

- The couple must demonstrate that they have a stable relationship, as well as report the number and sexes of the children they already have.

- Because PGD would only be used for family balancing, the couple must already have at least one child, and the sex chosen should be of the underrepresented sex.

- The couple would be told that sex selection is not completely accurate, and would be required to sign an agreement that a child of the unwanted sex would not be aborted.

To deal with more multifaceted issues and to rein in a potentially mushrooming business, Judith Daar, a professor of both law and medicine, has proposed three areas in which policies might be enacted. First, Daar posits that a license and certification could be required of providers, thus limiting the number of facilities offering sex-selection services. Second, informed consent should be required on the consumer's part, and third, criminal penalties would be imposed on those consumers and facilities who do not comply with licensing and consent regulations (Remaley 2000).

In this same vein, the Recombinant DNA Advisory Committee (RAC), created in 1974 to advise the NIH on matters of human gene technology, might be given accreditation and licensing powers to regulate the spread of prenatal diagnosis technology (Roberts 2002). The RAC would thus become analogous to the Human Fertilisation and Embryology Association (HFEA) in Britain, a governmental group that has the power to enact laws. It was the HFEA that in 1993 banned prenatal diagnosis for nonmedical reasons in Britain (Ethics Committee Report 2001).

Indeed, an alternative to regulating sex selection is to ban it completely. However, some feel a blanket prohibition would prevent therapeutic sex selection, would create a dangerous black market for the technology, and might be more difficult to enforce than carefully thought out incentives and regulations. In addition, policy regarding sex selection cannot be limited to simple prohibition or permission. Questions of whether it would be covered by health insurance or whether those who morally oppose it would be exempt from paying for any mandatory insurance coverage also come into play (President's Council on Bioethics 2003).

In Conclusion

There are many questions yet to be answered regarding sex selection. Is it justifiable to choose the sex of your child simply because you want to? Does permitting control over sex determination open the door to eugenics and the control over traits such as height, eye and hair color, and even intelligence?

Sex selection by infanticide and abortion has been common in certain areas of the world for centuries; the resulting changes in sex ratios, known as the gender gap, have become apparent in some regions. It is doubtful that the cultural pressures present in most Western societies would lead to the same type of stratification, but other questions arise. For example, how does sex selection influence the parent-child relationship or the expectations that parents hold for their children? What political, economic, and social repercussions would widespread sex selection have? These and other questions lead to amorphous, unanswered gray areas. However, the recent advent of powerful technologies that will almost certainly become more and more available in a free market economy attest that, for better or worse, we may be on the verge of obtaining concrete answers.

UNIT 4

Should We Allow Humans To Be Cloned?

Cloning may be good and it may be bad.
Probably it's a bit of both. The question must not be greeted
with reflex hysteria but decided quietly, soberly, and on its
own merits. We need less emotion and more thought.

RICHARD DAWKINS (1997)

CHAPTER 7

The Science of Cloning

The word "clone" comes from the Greek word *klon,* meaning a plant cutting or sprig. A complete plant may be propagated from a single piece of a parent plant; apple trees, for instance, are routinely grown from stem cuttings, and even a small part of the "runner" of a strawberry plant can eventually produce a whole field of fruit. Cloning is easy in plants, because they do not distinguish between their germ (reproductive) cells and somatic (body) cells until late in development. Nearly every plant cell can give rise to an entire new and fully functional plant that in turn is capable of reproducing. In some animal species (such as flatworms), this can also be done. However, such easy cloning is not the case in vertebrates.

The science of cloning animals is based on two principles. The first is that every nucleus in the embryo or adult carries the same genes (i.e., the genome established at fertilization). The second is that an egg can be "tricked" into normal cell division and development by procedures other than sperm entry. Remember that, as described in Chapter 3, fertilization carries out two processes: the transfer of genetic material, and the activation of development. These two events must be done artificially if cloning is to be successful.

Early Cloning Experiments: Sea Urchins and Amphibians

The idea for cloning began with the early experimental evidence that the genome is the same in all the somatic (body) cells. Before these experiments in the late 1800s, it was not known whether every cell in the body had the same genes, or whether each cell type differed by having only certain genes and not others. In one of the most important experiments in embryology,

Hans Driesch (1892) managed to separate the first four blastomeres of a sea urchin embryo (Figure 7.1). Each of the four cells was able to produce an entire new embryo, indicating that the nucleus of each cell contained the same genetic instructions and that each cell was **totipotent**—capable of forming all of the different cell types. Driesch's further experiments with 8- and 16-cell sea urchin embryos reinforced the idea that every nucleus in these embryos had all the genes needed to form any cell type, and in the correct pattern, to produce a viable adult.

In 1918, Hans Spemann showed the equivalence and totipotency of early vertebrate cells by constricting a fertilized salamander's egg with a thin loop made from a baby's hair. Only one side of the constricted egg contained a nucleus, and the first cell divisions took place only in the half with the nucle-

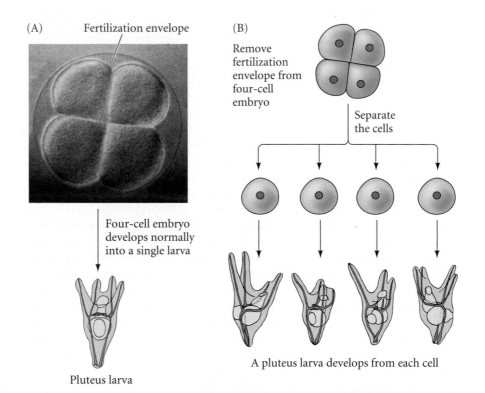

FIGURE 7.1   Driesch's experiment: totipotency of sea uchin blastomeres. (A) An intact 4-cell sea urchin embryo generates a normal sea urchin larva (pluteus). (B) When one removes the 4-cell embryo from its fertilization envelope and isolates each of the four blastomeres, each cell can form a smaller, but normal, pluteus larva. (All larvae are drawn to the same scale.) Note that the four larvae derived in this way are not identical, despite their ability to generate all the necessary cell types. (Photograph courtesy of G. Watchmaker.)

(A)

Jelly capsule of egg

Ligature

(B) Ligature relaxed; one nucleus allowed to migrate

(C)

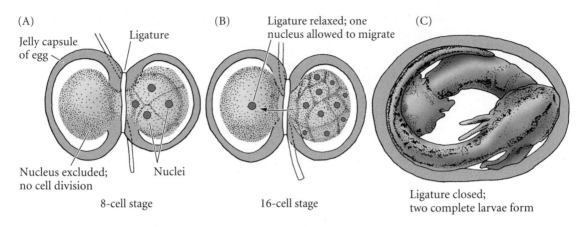

Nucleus excluded; no cell division

Nuclei

8-cell stage

16-cell stage

Ligature closed; two complete larvae form

FIGURE 7.2   Spemann's experiment: nuclear equivalence in newt cleavage. (A) When the fertilized egg of the salamander *Triturus taeniatus* was constricted with a ligature, the nucleus was restricted to one-half of the embryo. Cleavage on that side of the embryo reached the 8-cell stage, while the other side remained undivided. (B) At the 16-cell stage, a single nucleus entered the as-yet undivided half, and the ligature was constricted to complete the separation of the two halves. (C) After 140 days, each side had developed into a normal larva. (After Spemann 1938.)

us (Figure 7.2). Then, at the point when the cells were dividing such that the 8-cell stage was becoming the 16-cell stage, Spemann loosened the constriction and allowed a single nucleus to enter the vacant half of the egg. After the nucleus had entered, he fully constricted the hair loop so that the two halves were fully separated. One embryo had 15 cells, each with its own nucleus; the other embryo was a single cell containing the single nucleus that had been allowed to travel across the constriction. Both embryos developed into normal salamander larvae. Thus, each of the 16 nuclei must have contained all the genes needed to form the entire body of the amphibian.

All of these early experiments demonstrating totipotency were done on very early embryos in which the cells had not yet differentiated into the different cell types that make up an adult. In 1938, Spemann wrote that to test the idea that the genetic instructions contained in cell nuclei were identical, even after the cells have differentiated into gut cells or nerve cells or skin cells, one would have to have laboratory procedures to (1) remove the nucleus from an unfertilized egg, (2) place into this **enucleate egg** a nucleus taken from a somatic cell (i.e., a differentiated body cell that was neither an egg or sperm) and (3) activate development to see if that differentiated somatic cell nucleus could direct the forma-

> If without any deterioration, the egg nucleus could be replaced by the nucleus of an ordinary embryonic cell, we should probably see this egg developing without changes.
>
> YVES DELAGE (1895)

tion of all the cells in the body. Such procedures were eventually developed during the 1950s.

The restriction of nuclear potency

The ultimate test of whether the nucleus of a differentiated cell has undergone irreversible functional restriction is to have that nucleus generate every other type of differentiated cell in the body. If each cell's nucleus is identical to the zygote nucleus, then each cell's nucleus should be capable of directing the entire development of the organism when transplanted into an activated enucleated egg. Before such an experiment could be done, however, three techniques for transplanting somatic cell nuclei into enucleated eggs had to be perfected: (1) a method for enucleating host eggs without destroying them; (2) a method for isolating intact donor nuclei from somatic cells; and (3) a method for transferring the donor nucleus into the host egg without damaging either the nucleus or the egg. In addition, the egg would have to be appropriately activated in order for development to take place.

In the 1950s, the laboratory of Robert Briggs and Thomas King achieved all these techniques in experiments using leopard frogs (*Rana pipiens*). Briggs and King demonstrated that blastula cell nuclei could direct the development of complete tadpoles when transferred into the enucleated oocyte cytoplasm. Frog blastulas contain hundreds of cells (Figure 7.3), but the cells are not yet determined to be any particular type of cell. (Each cell of the frog blastula is therefore similar to the inner cell mass blastomeres of mammals; see Figure 1.8F.)

First Briggs and King combined the enucleation of the host egg with its activation. When an oocyte (immature egg) from the frog *Rana pipiens* is pricked with a clean glass needle (bursting the cortical granules and causing the release of calcium ions stored inside the egg; see Chapter 3), the egg undergoes all the cellular and biochemical changes associated with fertilization. The internal cytoplasmic rearrangements of fertilization occur, and the completion of meiosis takes place near the upper end of the egg. The meiotic spindle can easily be located at the upper pole, and puncturing the oocyte at this site causes the spindle and its chromosomes to flow outside the egg (Figure 7.4A). The host egg is now considered both activated (the fertilization reactions necessary to initiate development have been completed) and enucleated (all genetic material has been removed with the spindle and chromosomes).

The transfer of a new nucleus into the "empty" egg is accomplished by opening a donor blastula cell and

FIGURE 7.3   A frog blastula. The blastula contains hundreds of genetically identical cells that are not yet determined to be any specific type of cell. (Photograph courtesy of M. Danilchik and K. Ray.)

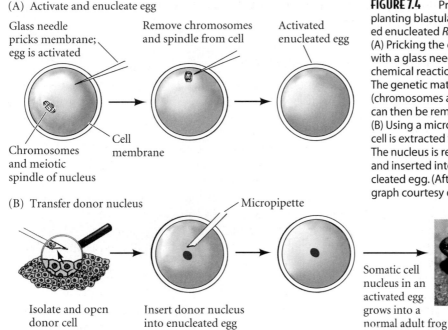

(A) Activate and enucleate egg

Glass needle
pricks membrane;
egg is activated

Remove chromosomes
and spindle from cell

Activated
enucleated egg

Chromosomes
and meiotic
spindle of nucleus

Cell
membrane

(B) Transfer donor nucleus

Micropipette

Isolate and open
donor cell

Insert donor nucleus
into enucleated egg

Somatic cell
nucleus in an
activated egg
grows into a
normal adult frog

FIGURE 7.4 Procedure for transplanting blastula nuclei into activated enucleated *Rana pipiens* eggs. (A) Pricking the egg's membrane with a glass needle triggers the biochemical reactions of egg activation. The genetic material of the nucleus (chromosomes and meiotic spindle) can then be removed from the egg. (B) Using a micropipette, a somatic cell is extracted from donor tissue. The nucleus is removed from this cell and inserted into the activated enucleated egg. (After King 1966; photograph courtesy of M. DiBerardino.)

transferring the released nucleus into the oocyte through a micropipette (Figure 7.4B). This is called **somatic cell nuclear transfer (SCNT)**, or more commonly, **cloning**. Some cytoplasm accompanies the nucleus to its new home, but the ratio of donor cell cytoplasm to recipient egg cytoplasm is miniscule—only 1 to 100,000—and the donor cytoplasm does not seem to affect the outcome of the experiments. With practice, this procedure produces live frog larvae (swimming tadpoles) about 80 percent of the time.

What happens when cell nuclei from more advanced embryonic stages are transferred into activated enucleated oocytes? King and Briggs (1956) found that whereas most blastula nuclei could produce entire tadpoles, there was a dramatic decrease in the ability of nuclei from later stages to direct development to the tadpole stage (Figure 7.5). When nuclei taken from the somatic cells of tailbud or later stage tadpoles were used as donors, normal development did not occur. Thus, most somatic cells appeared to lose their ability to direct development as they became determined and differentiated.

The pluripotency of somatic cells

Is it possible that some differentiated cell nuclei differ from others in their ability to direct development? John Gurdon and his colleagues, using slightly different methods of nuclear transplantation and a different frog

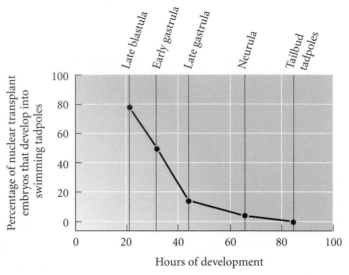

FIGURE 7.5   Percentage of successful somatic nuclear transplants as a function of the developmental stage of the donor nucleus in frogs (*Rana pipiens*). Once a larva reaches the "tailbud tadpole" stage (about 85 hours into its development), the nuclei from its somatic cells have differentiated and are no longer capable of directing development of a new individual. (After McKinnell 1978.)

species (*Xenopus laevis*, an evolutionarily more primitive frog than *Rana pipiens*), obtained results suggesting that the nuclei of some differentiated cells can remain totipotent (Figure 7.6). Gurdon, too, found a progressive loss of potency with increasing developmental age, although *Xenopus* cells retained their potencies for a longer period than did the cells of *Rana*.

To clone amphibians from the nuclei of cells known to be differentiated, Gurdon and his colleagues cultured differentiated epithelial (skin) cells from adult frog foot webbing. They knew the cells were differentiated because each of them contained a characteristic protein of adult skin cells. When nuclei from these skin cells were transferred into activated, enucleated *Xenopus* oocytes, none of the first-generation transfers progressed further than the formation of the neural tube shortly after gastrulation (see Chapter 1). However, through serial transplantation—taking nuclei from the cleaving cells of the cloned embryos and putting them into a second set of enucleated and activated oocytes—numerous tadpoles were generated. Although these tadpoles all died prior to feeding, they showed that a single differentiated cell nucleus still retained incredible potencies. A nucleus of a skin cell could produce all the cells of the young tadpole, even though it could not produce

Wild-type donor
of enucleated eggs
("surrogate mother")

Albino parents
of nucleus donor

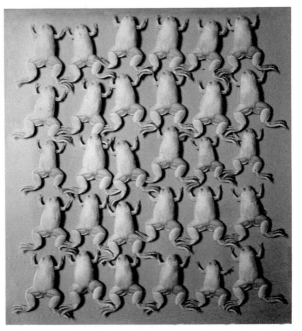

Cloned individuals identical to nucleus donor (female, albino)

FIGURE 7.6   A clone of *Xenopus laevis* frogs. The nuclei for all the members of this clone (right) came from a single larval individual—a female tailbud-stage tadpole whose parents were both marked by defective pigment genes (meaning they are albino, lacking skin pigmentation). Nuclei containing these defective pigment genes were transferred into activated enucleated eggs from a wild-type (a normal, non-albino) female. The resulting adult frogs were all female and albino, meaning that they are clones of the female tadpole whose nucleus supplied the genetic material. (Photographs courtesy of J. Gurdon.)

an adult. Rather than being totipotent, the differentiated nucleus is **pluripotent**: capable of producing many, but not all, cell types.

In the Briggs and King experiments, transplanted nuclei from larval cells produced adults, while Gurdon's work showed that transplanted adult nuclei could produce larvae. However, contrary to numerous science fiction books and magazine articles, a nucleus from an adult animal's differentiated cells had never produced another adult animal—until 1997, and the arrival of Dolly.

Cloning Mammals

Early in 1997, Ian Wilmut of the Roslin Institute in Edinburgh, Scotland shocked much of the world when he announced that a female Dorset sheep named Dolly, born to a surrogate mother in July 1996, had in fact been cloned from a somatic cell nucleus taken from an adult female sheep. This was the first time that an adult vertebrate had been successfully cloned using an adult nucleus, an event most biologists had predicted was years away from happening, if indeed it ever proved possible at all.

How did Wilmut and his colleagues "achieve the impossible"? First they took cells from the mammary gland (udder) of a pregnant 6-year-old Dorset

Why Clone Mammals?

Why do we need to clone mammals? Many of the reasons are medical and commercial, and there are good reasons why the techniques for mammalian cloning were developed first by pharmaceutical companies rather than at universities. Cloning is of interest to some developmental biologists who study the relationships between the nucleus and cytoplasm during fertilization, or by scientists who study aging (and the loss of nuclear potency that appears to accompany it), but cloned mammals are of special interest to those concerned with **protein pharmaceuticals**.

Important protein drugs include insulin, protease inhibitor (used in cancer therapies), and blood clotting factors. These drugs are difficult to manufacture biochemically. Some of them can be obtained from animals (insulin, for example, was traditionally obtained from pigs), but because of immunological rejection problems, patients usually tolerate human proteins much better than proteins from other animals. So how do we obtain large amounts of the specific human proteins we need?

One of the most efficient ways of producing protein pharmaceuticals is to insert the human genes that code for the desired protein into the oocyte DNA of sheep, goats, or cows. Such an insertion, achieved using recombinant DNA techniques, results in a **transgene**, and the animals containing such gene insertions are called **transgenic animals**. A transgenic female sheep or cow might not only contain the gene for the human protein, but might also be able to express the gene in her mammary tissue and thereby secrete the protein in her milk, as shown in the figure.

Producing transgenic sheep, cows, or goats is a highly inefficient undertaking. Only about 20% of the treated eggs sur-

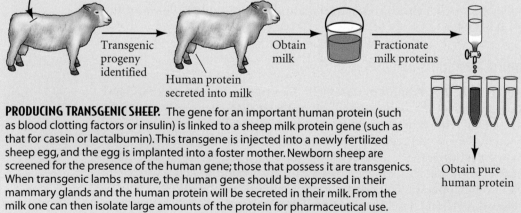

Gene for human protein

Sheep lactoglobulin gene

Sheep egg

Create transgene by recombinant DNA techniques

Inject transgene into nucleus of sheep's egg

Holding pipette

Implant egg into foster mother

Transgenic progeny identified

Human protein secreted into milk

Obtain milk

Fractionate milk proteins

Obtain pure human protein

PRODUCING TRANSGENIC SHEEP. The gene for an important human protein (such as blood clotting factors or insulin) is linked to a sheep milk protein gene (such as that for casein or lactalbumin). This transgene is injected into a newly fertilized sheep egg, and the egg is implanted into a foster mother. Newborn sheep are screened for the presence of the human gene; those that possess it are transgenics. When transgenic lambs mature, the human gene should be expressed in their mammary glands and the human protein will be secreted in their milk. From the milk one can then isolate large amounts of the protein for pharmaceutical use.

Why Clone Mammals? (continued)

vive the insertion technique and develop into transgenic adult animals. Of these adult transgenics, only about 5% actually express the human gene. And of those transgenic animals expressing the human gene, only half are female, and only a small percentage of these actually secrete a high level of the protein into their milk (plus it often takes years for them to first produce milk). Moreover, after several years of milk production, they die, and their offspring are usually not as good at secreting the human protein as the original transgenic animal.

Cloning would enable pharmaceutical companies to make numerous "copies" of an "elite transgenic animal." Such cloned transgenics should all produce high yields of the human protein in their milk. The medical importance of such a technology would be great, and human protein pharmaceuticals could become much cheaper for the patients, many of whom depend on them for survival. The economic incentives for cloning are enormous. Thus, shortly after the announcement of Dolly, the same laboratory announced the birth of a cloned sheep named Polly. Polly was cloned from transgenic fetal sheep somatic cells that contained the gene for human clotting factor IX, a gene whose function is deficient in hereditary hemophilia.

ewe and grew these cells in plastic dishes. The growth medium was formulated to keep the nuclei in these cells at a specific point (the "intact diploid stage") of the cell cycle; this cell-cycle dependence turned out to be critical. They then obtained oocytes (the maturing egg cell) from females of a different breed of sheep (Scottish blackface) and removed their nuclei. These oocytes had to be in the middle of their second meiotic division (which is the stage at which oocytes are usually fertilized).

The donor cells and the enucleated oocytes were fused by squeezing them together and sending electrical pulses through them; the electric pulses destabilized the cell membranes and allowed the cells to fuse together. Moreover, these same electric pulses activated the eggs to begin development (see Chapter 1). The cells divided, and the resulting embryos were eventually transferred into the uteri of pregnant Scottish blackface sheep. The procedure is diagrammed in Figure 7.7.

Of the 434 fused oocytes created during this experiment, only one survived to adulthood: Dolly. DNA analysis confirmed that the nuclei of Dolly's cells were indeed derived from the Dorset sheep from which the donor nucleus was taken. None of the genes necessary for development were lost or mutated in any way that would make them nonfunctional. That Dolly was a fully functional reproductive adult was proven when she mated normally with a male Dorset sheep and gave birth to her own offspring (Figure 7.8).

Dolly was euthanized in February of 2003. She was suffering from a progressive and intractable lung disease that is not uncommon among domestic sheep. She was 6 years old—middle age in sheep years. Whether being

FIGURE 7.7 Procedure used by Roslin Institute researchers to clone a sheep. Dolly, an adult Dorset sheep, was cloned by fusing a mammary gland (udder) cell nucleus from a Dorset sheep with an enucleated oocyte from a different breed of sheep, a Scottish blackface. The fused cells were then implanted into a surrogate mother (a Scottish blackface ewe), who gave birth to Dolly. (After Wilmut et al. 2000.)

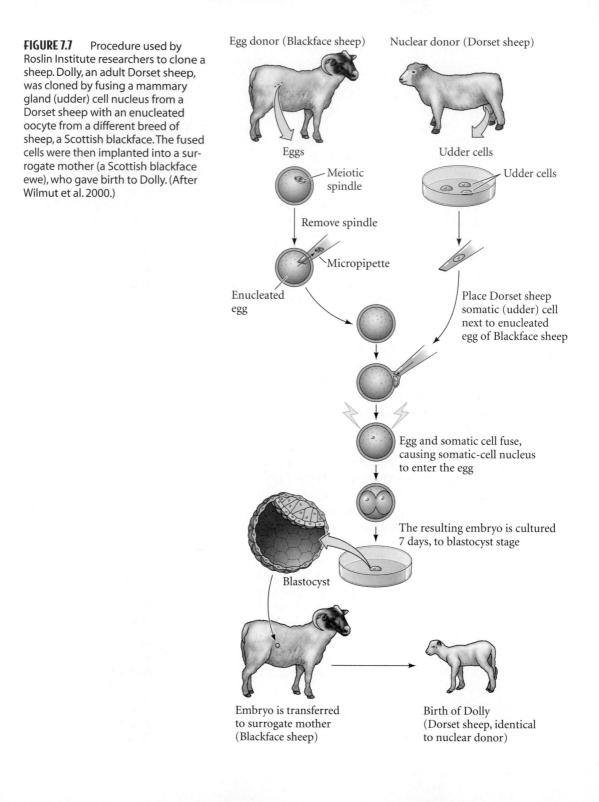

Egg donor (Blackface sheep)

Nuclear donor (Dorset sheep)

Eggs

Udder cells

Meiotic spindle

Udder cells

Remove spindle

Micropipette

Enucleated egg

Place Dorset sheep somatic (udder) cell next to enucleated egg of Blackface sheep

Egg and somatic cell fuse, causing somatic-cell nucleus to enter the egg

The resulting embryo is cultured 7 days, to blastocyst stage

Blastocyst

Embryo is transferred to surrogate mother (Blackface sheep)

Birth of Dolly (Dorset sheep, identical to nuclear donor)

FIGURE 7.8 The cloned sheep Dolly and her daughter. After she reached maturity, Dolly mated and produced a lamb (Bonnie, at right) by normal reproduction, thus confirming her condition as a fully functional adult mammal. (Photograph by Roddy Field, © The Roslin Institute.)

a clone contributed to her death at a relatively young age is still being debated.

Since 1997, laboratories around the world have achieved confirmed clonings of sheep, cows, mice, cats, and other mammals. Although it appears that all the organs were properly formed in the cloned animals, many of the clones developed debilitating diseases as they matured.

In addition, it is important to note that the phenotype (physical characteristics) of a cloned animal is not necessarily identical to the animal from which the nucleus was derived. The genotypes may be identical, but there is variability due to random events and due to the environment. The pigmentation of calico cats, for instance, is due to the random inactivation of one or the other X chromosome in each somatic cell of the female cat embryo. Therefore, the markings of the first cloned cat, a calico named "CC," were different from those of the calico cat whose cells provided the implanted nucleus that generated the clone (Figure 7.9). The same genotype gives rise to multiple phenotypes in cloned sheep as well, as Wilmut commented

[The four sheep] are genetically identical to one another and yet are very different in size and temperament, showing emphatically that an animal's genes do not "determine" every detail of its physique and personality.

IAN WILMUT, *THE SECOND CREATION* (2000)

FIGURE 7.9 The kitten "CC" (left) was cloned using somatic nuclear transfer from "Rainbow" (right). Their markings are not identical because the pigmentation of calico cats is affected by the random inactivation of one of the two X chromosomes in the female cells. The behaviors of these cats also differ. (Photographs courtesy of the College of Veterinary Medicine, Texas A&M University.)

Why Do Clones Have Health Problems?

The question of why so few cloned animals survive to be born and why those that do survive tend to have serious health problems may be related. One theory is simply that cloned animals may be prematurely old; that is, the newborn clone's cells may already reflect the age of the adult animal from which it was cloned.

One mark of cellular aging is shortened telomeres. **Telomeres** are small DNA sequences found at the end of normal chromosomes (Figure A). With each cycle of cell division, the telomeres shorten slightly, until eventually they are so short that the chromosome can no longer replicate.

When Dolly was 3 years old, for example, researchers found that her telomeres were the average length of those expected in a 6-year-old sheep—the age of the sheep from whose nuclei Dolly was cloned. When Dolly died at 6 years of age, some scientists believed that her biological age was in fact significantly older than her chronological age. This explanation does not hold with all species however. Cells in cloned mice have shown no signs of premature aging, and in another study calves cloned from a senescent cow (one whose own cells could no longer replicate) had telomeres that were no shorter than those of normal calves.

Another approach to the question of poor health among clones is based on the fact that the DNA of the differentiated adult cells used as nucleus donors is highly modified, and it may be extremely difficult to return the DNA to the unmodified, undifferentiated state found in the early blastomeres. Even though different cell types all contain the same complement of genes, it is clear that not every gene can be active (i.e., produce its protein) in every cell—the result would be molecular chaos.

The major type of DNA modification under scrutiny is **methylation**. In methylation, methyl groups—small biochemical groups made up of one carbon atom and three hydrogen atoms (CH_3)—are placed on the DNA molecule, preventing the gene from being activated. Different regions of DNA are methylated in different cell types: in red blood cell precursors, for example, the DNA of the globin genes is unmethylated (active), but the gene for insulin is methylated (inactive). In the pancreatic cells that secrete insulin, the methylation pattern is reversed (Figure B).

Several laboratories have found that the genes of cloned animals have abnormal methylation patterns. Apparently, while many of the genes can be "re-set" to their undifferentiated state (referred to as **epigenetic reprogramming**), other genes may retain their differentiated methylation pattern. Faulty methylation patterns leading to faulty gene activation (i.e., faulty protein production) would explain why so few cloned embryos sur-

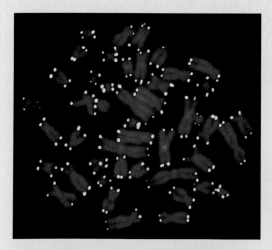

(A) TELOMERES ON HUMAN CHROMOSOMES. The DNA sequence that form telomeres at the tips of chromosomes shows up in fluorescent yellow in this micrograph of human chromosomes. (Photograph by Peter Lansdorp/Visuals Unlimited.)

Why Do Clones Have Health Problems? (continued)

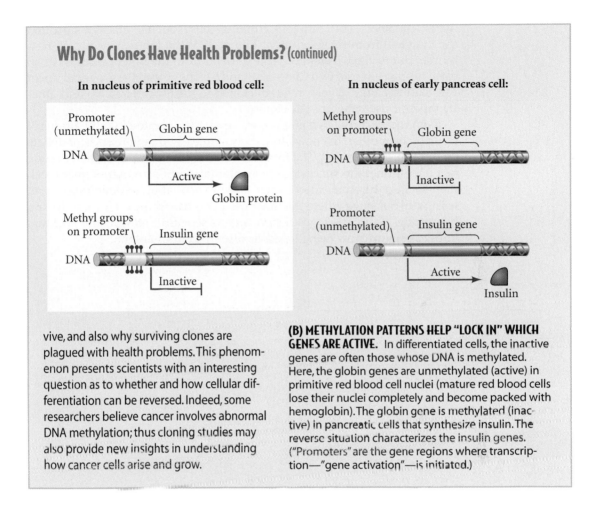

In nucleus of primitive red blood cell:

In nucleus of early pancreas cell:

vive, and also why surviving clones are plagued with health problems. This phenomenon presents scientists with an interesting question as to whether and how cellular differentiation can be reversed. Indeed, some researchers believe cancer involves abnormal DNA methylation; thus cloning studies may also provide new insights in understanding how cancer cells arise and grow.

(B) METHYLATION PATTERNS HELP "LOCK IN" WHICH GENES ARE ACTIVE. In differentiated cells, the inactive genes are often those whose DNA is methylated. Here, the globin genes are unmethylated (active) in primitive red blood cell nuclei (mature red blood cells lose their nuclei completely and become packed with hemoglobin). The globin gene is methylated (inactive) in pancreatic cells that synthesize insulin. The reverse situation characterizes the insulin genes. ("Promoters" are the gene regions where transcription—"gene activation"—is initiated.)

about the different size and personalities of four sheep cloned from blastocyst nuclei taken from the same embryo. Wilmut and other conclude that, for this and other reasons, the "resurrection" of lost loved ones by cloning is not feasible.

Cloning Humans

While a cloned human has yet to be realized, the first steps have already been performed. Hwang Woo-suk and Moon Shin-yong of Seoul National University in South Korea have successfully coaxed a line of totipotent embryonic stem cells to develop from human oocytes nucleated with human somatic cell nuclei (Hwang et al. 2004). Hwang, Moon, and their team created the human embryo after collecting 242 eggs from 16 anonymous volunteers. They also took cumulus cells (somatic cells that surround the oocytes) from these women. The goal of the lab was not to clone a human but to clone a stem cell line derived from the inner cell mass of an early human embryo (see Chapters 1 and 9). The researchers' overall intent was said to be the development of cloned tissue for transplants and other therapeutic uses (therapeutic cloning).

Even though no cloned embryo was ever transplanted into the uterus of a surrogate mother, the news from Seoul unleashed new rounds of debate on reproductive cloning. The first step toward human reproductive cloning has been done: the successful transplantation of a human somatic cell nucleus into the enucleated oocyte and the activation of development, at least until the stage where the inner cell mass has formed. Human cloning has left the realm of science fiction, and its scientific, ethical, and spiritual implications are now being faced, as discussed in Chapter 8.

CHAPTER 8

Ethics and Policies
for Human Cloning

One week after the news about a cloned sheep named Dolly was made public, a poll found that 90 percent of Americans thought that human cloning should be banned, 67 percent thought that the cloning of animals was unacceptable, and 56 percent would not eat the meat of a cloned animal (McLaren 2001). Some of this response might have been due to an explosion of sensational reporting by the American media. Some media sources even dug up old science fiction scenarios predicting the emergence of armies of drones taking over entire countries, clone farms used for spare parts, and evil dictators achieving immortality by cloning themselves for generations (Butler 1997; Klotzko 2001). Such reports calmed down after those first few weeks, but many people remain ethically opposed to cloning.

The questions arising from cloning (or, more accurately, from somatic cell nuclear transfer, or SCNT) in fact fall into two separate categories: therapeutic cloning and reproductive cloning (see Table 9.1 on page 153). Therapeutic cloning is the cloning of cells, not people or animals. It is proposed as a possible way to provide tissue transplants that are genetically identical to the patient receiving them. In this situation, the patient's genome would be transferred to an ovum and stimulated to start dividing. After several days, cells from the inner cell mass—stem cells—would be removed from the blastocyst and grown in vitro to differentiate into the tissue required by the patient. The tissue could then be transplanted into the patient without fear of immune rejection, since the tissue would be genetically identical to the patient, and would be recognized as self. Although therapeutic cloning has many opponents, most scientists and medical researchers believe that its tremendous potential benefits outweigh its risks. Therapeutic cloning and stem cell research will be the subject of the next unit.

This chapter will concentrate mostly on the scientific and ethical concerns surrounding reproductive cloning. In reproductive cloning, an embryo produced in vitro (from the insertion of the subject's nuclear material into an enucleated oocyte) is implanted into the uterus of a surrogate mother and allowed to come to term. The result is an offspring whose genome is identical to that of its nuclear donor—in effect, an identical twin of the nuclear donor.* The reproductive cloning of human beings is a concept has met with significantly more resistance than that of therapeutic cloning. In addition to major health issues facing the embryo and child, many feel that such cloning may infringe on human dignity, individuality, and autonomy.

Why Should We Clone Humans?

There have been several arguments made that represent reproductive cloning as a moral, indeed, a responsible action under certain circumstances. Arguments in favor of human reproductive cloning include:

1. It offers everyone the ability to produce biologically related children. For instance, if a man could not produce sperm, cloning would allow him to have a biologically related child. (The child would not, of course, be biologically related to his partner.)

2. It offers the ability to avoid passing along certain genetic diseases. If both partners of a couple carry a recessive gene that puts them at high risk of producing a child with a seriously debilitating medical condition (such as cystic fibrosis or sickle-cell anemia), cloning would get around this problem.

3. It offers the ability to produce organs and tissues for transplantation. For a person needing, say, a bone marrow or kidney transplant, cloning would allow a genetically identical person to be made who could then serve as the donor.

4. It offers the ability to replace a loved one. A child killed in a car accident, for example, could be "replaced" by cloning a new infant from the dead child's cells.

5. It offers the ability to reproduce those whom society deems valuable, giving that society the opportunity to clone individuals who possess what are considered superior genetic endowments. This could include scholars, athletes, and creative artists.

---

*The use of the "identical twin" metaphor here is not exact. The clone would not have the same mitochondrial genes as its nuclear donor (because mitochondria come only from the egg), nor would it develop in the same uterine environment its donor developed in.

6. Freedom is better than prohibitions, and governments should not restrict freedom if no actual harm is documented. The desire for a biologically related child should not be restricted if no harm is documented.

In 2002, the President's Council on Bioethics identified these as being some of the major arguments that have been used to justify cloning. Some of these, such as the ability to avoid genetic disease, can be circumvented by less controversial procedures. Other arguments involve questionable assumptions, such as that a clone of a loved one will act the same as the original person even if raised in a totally different environment. We will be looking at some of these arguments and some proposed rebuttals to them.

In the United States, perhaps the principal argument in favor of allowing reproductive cloning research to go forward is the widely held belief that the government should not restrict freedom *when no actual harm is documented* (see Robertson 1994, 1998a). The major argument for restricting cloning research then becomes, simply, whether or not it is safe. Therefore, we begin with the issue of safety; for if the techniques cannot be made safe and efficient, it is doubtful human cloning will be attempted.

The Safety of Reproductive Cloning

The main scientific arguments against cloning cite the low success rate of mammalian cloning and take the stand that cloning is simply too dangerous even to consider performing on humans.

In considering the safety of any new or experimental procedure (as, for example, a clinical drug trial), it is important to bear in mind the matter of patient consent. Drug testing is done only with the informed consent of volunteers. Human reproductive cloning, on the other hand, would involve the creation of a new person, who could not be asked for their consent and who could potentially suffer major physical and/or psychological problems. Currently, concerns of safety for the embryos (as well as that of the many surrogate mothers who would be needed) are a strong argument against reproductive cloning.

Physical well-being and accelerated aging

Dolly died early in 2003 at the age of 6, half the average life expectancy of a healthy sheep. She died of a progressive lung disease that normally affects older sheep and expressed other signs of accelerated aging, such as obesity and arthritis. Mice cloned using the same method often die shortly after they are born, and many others die of liver disease and pneumonia before they reach the average lifespan of a normal mouse. In addition, most cloned mice develop in enlarged placentas, suffer from obesity, liver failure, brain malformations, respiratory distress, and dysfunction of the circulatory and immune systems (Jaenisch and Wilmut 2001; Ogonuki 2002; Humpherys et al. 2002). Some scientists are optimistic that these problems are rooted in technique rather than in developmental physiology, and that researchers will find ways to circumvent the problems. Others, however, have begun to study the genetics of clones and assert that the abnormalities are inherent in the very process of using a differentiated somatic cell to create a new organism.

The problem of premature aging may involve the telomeres, as described on page 122. During normal meiosis—the process by which germ cells are created—the parental DNA is rejuvenated and repaired, and telomere length is restored (McKinnell 1999; McLaren 2000). These processes are bypassed in cloning.

Epigenetic reprogramming and chromosome replication

In the meiotic processes of oogenesis and spermatogenesis, the genome of each germ cell is methylated in a specific manner, leading in turn to specific gene expression during development (see page 123). In humans, this **epigenetic reprogramming** takes months (in the case of spermatogenesis) or years (in the case of oogenesis). Even in the case of in vitro fertilization, which uses gametes that have undergone normal epigenetic reprogramming, the mere physical manipulation of the technique can lead to abnormal methylation in the embryonic genome (Khosla et al. 2001; Young 2001). In cloning, the reprogramming process is given only minutes to hours to take place (Jaenisch and Wilmut 2001), and a study at the University of Pennsylvania indicates that the resulting methylation patterns may very well be random (Boiani et al. 2002).

A leading scientist in the field, Dr. Rudolph Jaenisch of the Whitehead Institute, asserts that between 30 and 50 percent of genes are improperly expressed in a developing clone.* There is currently no technology able to determine the epigenetic state of a cloned embryo, and hence no possibility for "quality control" in selecting which embryos to implant (Jaenisch and Wilmut 2001).

Attempts to clone primates such as humans may be hampered by another block as well. Recent experiments using rhesus monkeys have indicated that nuclear transplantation repeatedly fails in the monkeys, even under exactly the same conditions it succeeds in mice (Simerly et al. 2003). Primate somatic-cell nuclei may have difficulty making the mitotic spindles needed for replication when placed into oocytes. Such a failure could be an insurmountable block to cell division, and may present a serious obstacle to human reproductive cloning.

---

*Although this faulty methylation of genes may have dramatic negative effects on a cloned organism, it appears not to affect tissues created in therapeutic cloning (Jaenisch, quoted in Hall 2004).

Playing the numbers

In the cloning procedure that led to the birth of Dolly, Ian Wilmut and his colleagues started with 277 fused oocytes, of which 29 developed enough to be implanted into 13 different ewes. Of those 29 embryos, only Dolly achieved full-term development and birth. It took more than 150 embryos to achieve the first cloned mouse pup, and 586 embryos to get the first two cloned piglets. Based on this work, Wilmut estimates that it would take an average of 1,000 human eggs implanted into 50 different women to produce a single human cloned offspring (Klotzko 2001).

Unlike therapeutic cloning, in which stem cell lines are grown in vitro, reproductive cloning requires a surrogate mother. Using Wilmut's calculations, the lives of at least 50 women would be interrupted in attempting to have a baby for someone else. Currently, when a surrogate mother agrees to carry a baby for a couple, she is carefully chosen by the prospective parents and often plays an integral role in raising the child. Being a surrogate mother is seen as a huge responsibility. To require 50 surrogate mothers to produce one child is a phenomenal undertaking (and expense). Furthermore, the result of one baby per 50 implanted cloned embryos is merely a statistical prediction. What if in fact more than one of the clones survives to be born? Who is then responsible for the lifelong care of the "extras"?

Scientists weigh the risks

Many scientists feel that these problems, which have no foreseeable solutions, are more than enough reason to ban human reproductive cloning. Two of the pioneering leaders of the field, Ian Wilmut and Rudolph Jaenisch, have pointed out that if human reproductive cloning is attempted and results in horrific health problems, the public may be shocked into banning all forms of human cloning—including therapeutic cloning, which these men believe can greatly advance medical research and save thousands of lives.

However, there are researchers who feel that at some point the biological difficulties of reproductive cloning will be smoothed out, and that the scientific advances made along the way will be worth the risk and effort. Some advocates of cloning argue that we will discover a tremendous amount about human genetics and development that we would never learn otherwise. They assert that issues of safety alone are not sufficient to justify banning reproductive cloning completely (see McLaren 2001).

Proponents claim that if complete safety were the criterion for legitimacy, medical progress would come to a screeching halt. Testing new drugs or surgical procedures can be dangerous, but the benefits far outweigh the risks. However, in following this analogy, advocates of banning cloning liken cloning to a drug that kills most of the animals on which it is tested (or at the least leads to their premature death). Would a drug with such results be cleared for human testing? Of course not. But in fact this is what many are calling for in the case of reproductive cloning.

Can cloning offer benefits that would eclipse the enormous risk? At this point, proposed applications of cloning include the prevention of genetic disorders of which both parents are carriers, and the circumvention of mitochondrial diseases—serious conditions resulting from defects in mitochondrial DNA, which comes only from the mother—by using the egg of a woman other than the mother (Seidel 2000; Gillon 2001). Both benefits can already be attained through IVF and adoption.

Ethical Positions on Human Cloning

It is possible that the safety issues surrounding cloning may be overcome in the future, and other issues of ethics will be relied on to decide whether cloning is morally acceptable. The idea of manufacturing embryos strikes an ill chord with many people, and cloning becomes immoral simply because it is unnatural. Others attack this argument as invalid because much of what humans do—from flying in airplanes to the hybridization and domestication of plants and animals to advanced medical procedures—could be considered unnatural (Steinbock 2000). Blood transfusions, mechanical life support systems, feeding tubes, organ transplants, open-heart surgery, in vitro fertilization, and defibrillators that use electricity to re-start a stopped heart are all "unnatural," but they are not banned and the majority of people find them morally acceptable.

One of the major contentions raised by opponents of cloning is that it is ultimately a pernicious process, involving the destruction of hundreds of embryos. Much of this argument is associated with differing ideas of when human life begins, and the religious and ethical views discussed in Chapter 2 are applicable. However, the nature of some of the questions raised by cloning—indeed, it can be seen as "playing God"—have prompted philosophers and theologians to express opinions on the subject.

Religious views

While each of the world's major religions has something to say about cloning, there is very little agreement among or even within the various religions as to whether cloning technology is ethical. Theologians must infer and deal with new ground, since the texts on which the world's major religions are based—including the Old and New Testaments, the Talmud, the Qur'an, and the Vedas—naturally cannot specifically address questions arising from a technology that is less than a decade old.

JUDAISM   Medicine and healing are of great importance in the Jewish tradition, and for this reason therapeutic cloning finds huge support among

Jewish communities worldwide. Reproductive cloning, on the other hand, raises a much more complex set of issues. Some Jewish scholars say there are no moral or legal objections in a religious sense, while others bring up reasons why the technology should be condemned.

According to Jewish tradition, humans are partners and co-creators with God, and it is not only permissible, it is our duty to intervene in nature to alleviate suffering, especially if medical advances are the result. Avraham Steinberg, director of the Center for Medical Ethics at Hebrew University-Hadassah Medical School, raises the argument that having children for purposes other than fulfilling the mandate of "be fruitful and multiply" is not wrong. He states that bearing a child for the purpose of organ donation is in fact two good deeds: bearing one child, and saving another (Wahrman 2002). However, Steinberg also brings up that God's mandate was "be fruitful and multiply," not "replicate yourselves," and therefore cloning may not in fact be God's intent for humans.

> The Jewish tradition advises one to pause before one permits that which can lead down a variety of slippery slopes whose consequences we do not fully understand, and whose results we cannot predict.
>
> RABBI MICHAEL J. BROYDE (2002)

One of the major issues brought to the table is the concern that cloning would disrupt the traditional relationships among family members, since the child would have no identifiable parenthood under *halakhic* law (*halakha* refers to Jewish civil and family law, derived from the religious precepts of the Torah and Talmud). However, because cloning does not involve a sex act, it is not expressly forbidden by the laws dealing with husbands and wives (Broyde 2002; Wahrman 2002).

Among other concerns raised by Judaism are the risk of birth defects and the transmission of somatic mutations associated with cloning. Even though fetuses are not awarded human status until birth, some feel that the level of embryo abnormalities involved in cloning is unacceptable in both the religious and secular worlds. Judaism does not, however, worry that clones will be soul-less (Wahrman 2002).

The Israeli Knesset (parliament) has placed a 5-year moratorium on the technology, and the Union of Orthodox Jewish Congregations of America and the Rabbinical Council of America both oppose reproductive cloning, while endorsing therapeutic cloning (Broyde 2002; Wahrman 2002).

CHRISTIANITY The Roman Catholic position is straightforward and objects to human cloning on the same grounds that fuel its opposition to assisted reproduction, birth control, and abortion: (1) these technologies uncouple the acts of marital intercourse and procreation; and (2) because Catholics believe that life begins at conception, the destruction of embryos required by these procedures is seen as the equivalent of abortion. The United States

Council of Catholic Bishops is on record as opposing both reproductive and therapeutic cloning (USCCB 1998).

Mainstream Protestant Christianity has by and large urged a moratorium on cloning research until the ethics of the situation can be explored at length. However, it has no inherent objection to reproductive cloning, as long as the human dignity of the clone is assured. Evangelical Christianity contends that human cloning deprives the clone of inherent humanity in an ungodly creation. In addition, Evangelical Christians often believe that human cloning would place us on a "slippery slope" leading to a revival of eugenics, where only certain people have the right to reproduce (Campbell 1997).

ISLAM The Qur'an, while supporting human intervention for the betterment of human health, warns against human arrogance in such situations. In addition, the Qur'an considers sex pairing, which cloning eliminates, "to be the universal law of all things." Many Muslims object to human cloning on the grounds that clones would experience discrimination, and could potentially become commodities rather than individuals (Campbell 1997).

At the Ninth Meeting of the Fiqh-Medical Seminar, the Islamic Organization for Medical Sciences (IOMS) discussed the risks of cloning and recommended to the Islamic community that human nuclear transfer be banned (Islamic Views 2004). Similarly, the Fatwa Council, the highest Muslim authority in Palestinian territories, has banned all forms of human cloning.

However, Dr. Abdulaziz Sachedina, a professor of Islamic Studies at the University of Virginia, has scrutinized both cloning technology and Islam to find precisely where the two harmonize and conflict (Sachedina 2004). He emphasizes that Muslims see humans as co-creators with God (as do the Jews), and that we are actually encouraged to interfere with natural processes when improvement of human well-being is the goal. In addition, since Muslims believe that ensoulment of the embryo occurs long after fertilization (see Chapter 2), manipulation of pre-implantation embryos is not forbidden. Finally, because cloning has been recommended to confer fertility within the boundaries of marriage, it should not be opposed. He concludes that if cloning causes no harm, and it can be kept from political corruption, it should not be denounced by Islam.

HINDUISM In June of 1997 the Hindu news magazine *Hinduism Today* was asked to submit an official statement to then-President Bill Clinton regarding the Hindu view on cloning. The statement (Palaniswami 1997) noted that there is a widespread aversion to cloning among Hindus, and that many religious leaders feel that cloning is arrogant and sinful in that it is playing God. Even those who do not deem cloning sinful believe that it is unnecessary and not worth the immense risk. The fact that a person's clone may live past the original nuclear donor's death raises difficult questions regarding *karma* and the path of the soul after death. In their recommendations to President Clinton, *Hinduism Today* called for careful regulation of the technology and the involvement of spiritually minded people in the legislative process. The publication notes that Hindus neither condemn nor condone the progress of science, but point out that it needs to be done with good intentions: "if done in the service of selfishness, greed and power, it may bring severe karmic consequences" (Palaniswami 1997).

BUDDHISM Buddhists have no official policy toward human cloning, since each person is led separately by the teachings of the Buddha. Cloning itself is not seen as inherently wrong. Of the two main components required for forming life, *rupa* (matter) and *nama* (mentality), cloning only involves manipulation of matter while mentality is left untouched to develop on its own (Kirkpatrick 1999). Pinit Ratanakul of Mahidor University, Thailand states his position that, in general, cloning would pose no ethical problems for Buddhists as long as no one was harmed (Ratanakul 2000).

> Could we be sure of the rightness of the desire to experiment with human cloning? Could this be the product of our selfish desire to perpetuate ourselves by interfering with natural reproduction, and by indulging in self-delusion? Is it an attempt at bending nature to conform to our will, our self-image?
>
> PINIT RATANAKUL (2000)

Although Buddhists are aware that cloning might have some benefits, many are wary of the technology. On the one hand, it would allow more chances to reach enlightenment (Campbell 1997). On the other hand, Buddhists worry about ensuring the moral righteousness of cloners' motivations. They also fear that tampering with the natural order in such a profound way could lead to damage greater than can be fixed (Ratanakul 2000). Hwang Woo-suk of Korea, a Buddhist and one of scientists credited with cloning the first human embryo, opposes reproductive cloning, but feels that, in some ways, therapeutic cloning "restarts the circle of life" (Dreifus 2004).

Cloning and cultural values

As we saw in Chapter 7 (see page 118), one proposed use of reproductive cloning is to clone farm animals whose abilities in a specific area (such as milk production) make them superior (for our purposes) to other individuals. In this vein, some have expressed the fear that widespread human reproductive cloning would open the door to eugenics– producing more people with "desirable" qualities (see Chapter 12). But who decides what qualities are desirable? It doesn't take much of a stretch of the imagination to see where such reasoning could lead.

Lori Andrews relates attending a discussion of cloning organized in the Arab Emirate of Dubai, during the course of which it was agreed upon that cloning humans is not forbidden *if the clone is that of a married man*. This brought her to one of her conclusions about the ethics of cloning: "It was then that I realized that somatic cell nuclear transfer … was not so much a scientific technique to reproduce individuals, as a way to clone our values. The Raëlians* would clone smart people, because that was what they valued. The Muslims would clone men" (Andrews 1999).

Extending these concerns about eugenics, some have even worried that widespread cloning could essentially halt human evolution (Butler 1997). They fear that the beneficial recombination of genes through traditional gamete formation and fertilization will be lost because everyone will be reproducing asexually. Counterarguments point out that this is a very unlikely scenario; cloning is an expensive technology, and could never be widespread enough to significantly alter population genetics (McLaren 2001; Singer 2001). In addition, the variety of individual preferences should maintain a variety of traits among clones. There is also the contention that humans have already distanced themselves from evolution through modern medicine and technology, and thus cloning would have no significant impact beyond what is already in place (McLaren 2001).

Cloning and psychological well-being

Ethical concerns regarding the psychological well-being and integrity of both cloned offspring and those who create them play a huge role in whether human reproductive cloning should be considered permissible. Some argue that to ban cloning would be to infringe on the reproductive rights of the parents (Burley 1999; Robertson 1998a,b). Others point out that in fact people do *not* have the right to reproduce however they want to, especially when great harm may potentially befall a child thus conceived (Gilbert 1998; Human Genetics Alert Campaign 2002). Indeed, virtually no one would argue with such societal regulations: one cannot legally have sex with one's parent, child, or sibling, or with minor children. A ban on reproductive cloning is not a ban on reproduction, but would merely prevent one expensive and dangerous method of reproduction in which the child would be potentially psychologically scarred and physically unhealthy.

One argument points to the selfishness of utilizing this new reproductive technology. What benefit does it have for the child being created? According to Kantian philosophy, children should be created as an end in themselves and not as a means to an end (Kitchen 2000; Gillon 2001). But wouldn't all cloned children be created as means to ends, because if the child were actually wanted as an end in itself, more conventional and much

---

*The Raëlians are a quasireligious organization with a keen interest in cloning. They are believed to promote a form of government they refer to as "geniocracy," or rule by geniuses. A company called Clonaid, apparently founded and run by the Raëlians, claims to have produced at least two living cloned babies but has never offered any scientific evidence for the claim, which is regarded as a hoax.

cheaper methods could be used (Lewontin 2000)? If a person is incapable of having children, in vitro fertilization, sperm donation, and adoption are all options that do not pose nearly the biological risks associated with reproductive cloning.

One proposed use for cloning is to create " donor matches" for relatives (usually siblings) who need organ or tissue donations (Robertson 1998b). Following Kantian logic, it would seem that creating a clone to save the life of another person is unethical, just as it would be unethical to try to "replace" a dead child by cloning it. It is creating a life to fulfill a task. What if that task is not completed? Will the child be seen as a failure? Furthermore, what psychological burdens would be placed on a child who knows he or she was created solely to save a sibling's life? What if the sibling dies? Will the clone be seen as, or feel like, a failure?

Bearing children so they can be used as matching organ or tissue donors is something that is already done through natural means.* Some would argue that cloning would merely allow this process to happen without playing the "genetic lottery," increasing the chances that a new baby will be able to successfully donate an organ (Singer 2001). Proponents maintain that there is no need to suppose that the new child will not be loved just as much as the first child, even if it does not succeed at its task. But many people are uneasy about creating new life to save an existing one. (Moreover, tissue donation for transplantation might be created through therapeutic cloning, which does not involve the creation of another person or involve psychological harm to a clone.)

In debating the production of children as a means to an end, ethicists discuss arguments about robbing children of their individuality. One could conclude that "individuality" is defined by personal memories and interpretations of the memories, and thus a clone is not at risk of losing this quality (Kitchen 2000). However, Daniel Brock, a professor of philosophy at Brown University, points out that a distinction must be made between an actual loss of individuality and the clone's perceived loss of individuality. He notes that losing one's *sense* of individuality is just as painful to a person as actually losing that individuality (Brock 2002).

Brock also asserts that ignorance of how one will turn out is necessary to a feeling of individuality and that knowledge about the nuclear donor's life and achievements might lessen a clone's sense of uniqueness. Brock and others warn that individuality is maintained partially in the opinions of others and that clones run a great risk of being judged on their genotype's

---

*Although needing "spare parts" may sound like a bad reason for having a child, one can ask what a *good* reason might be. Is having a child in order to save an existing child any worse than having a child to save a marriage, or to perpetuate your family's name? About half of all U.S. children are conceived "accidentally" (Nulman et al. 1997). We do not usually question the reasons why a child is conceived.

expected phenotype, a phenomenon called "living in the shadow" (Burley 1999; Holm 2001).

By "living in the shadow," a clone might be considered to be in danger of losing its individuality as characterized by the British philosopher and ethicist John Stuart Mill (1806–1873), who considered the individual's ability to freely choose their own path to be an essential component of human liberty.* It can be argued that any child may have their individuality "stolen" by overbearing parents (Mill himself was the child of a rigorously demanding and controlling father), but one can imagine that this risk might be much greater for a cloned offspring. If it is already hard for a child or teenager to live with a parent who constantly pushes them to achieve more—to do better academically, master a musical instrument, be more of a leader, or excel in sports—what might be the psychological outcome if the child were in fact a clone of that academically or athletically or artistically successful parent?

> No one's idea of excellence in conduct is that people should do absolutely nothing but copy one another.
>
> JOHN STUART MILL, *ON LIBERTY* (1859)

Public Opinion and Public Policy

A 2002 survey by the Pew Research Center showed that fewer than 25 percent of Americans supported human cloning. Most of those who objected believed that human cloning is fundamentally immoral; a few objected solely on the grounds that they did not feel cloning technology is safe enough yet (Pew 2002). Leon Kass, the head of the President's Council on Bioethics, has called the negative gut-level reaction many people have to the idea of human cloning the "wisdom of repugnance." He suggests that because this is such a widespread reaction, even if it is an unreasoned one, it should be considered at least potentially valid (Kass and Wilson 1998).

In the international arena, over 30 countries have banned reproductive cloning (Biever 2003). In the United Kingdom, the Human Fertilisation and Embryology Authority banned human reproductive cloning in 2001 (Koerner 2002; Janson-Smith 2002). The Genetics Policy Institute, a major international advocacy group supporting stem cell therapy (i.e., therapeutic cloning), also supports a ban on human reproductive cloning (GPI 2004). And the British Royal Society's task force on stem cells has presented a synthesis of the views of 67 scientific societies, all of which endorse research on therapeutic cloning while favoring a ban on reproductive cloning (Nature 2005).

---

*Mill, an economist, political theorist, government administrator, and essayist, was perhaps the most influential English-speaking philosopher of the nineteenth century (Wilson 2003). Many of his views on individual liberty are incorporated in the social and legal fabric of today's democratic Western societies, including (perhaps especially) that of the United States.

Shaping Policy in the U.S. Government

When government agencies in the United States consider creating new laws that involve scientific issues, there are several agencies they can call upon for advice. Foremost among these is the National Academy of Sciences, which was established by the United States Congress in 1863 to advise the federal government on scientific and technical matters. The National Academy of Sciences is composed of scientists, and it has convened panels to discuss many issues where science intersects with other societal interests. The National Academy of Sciences has issued several reports concerning stem cells and cloning. Its advice is, in sum, that reproductive cloning be banned (NAS 2001a), but that therapeutic cloning be permitted (NAS 2001b, 2002).

The National Academy of Sciences is part of a larger scientific advisory body that includes the National Academy of Engineering, the Institute of Medicine, and the National Research Council (NRC). This group, known simply as the National Academies, has reinforced the NAS position in a statement of policy guidelines for therapeutic cloning (see page 167).

Lawmakers can also choose to look for advice from the President's Council on Bioethics. The Council has no legislative ability, but it functions as an advisory agency to the President and can make policy recommendations. The Council's mandate is "to learn and advise the President on bioethical issues that may emerge as a consequence of advances in biomedical science and technology." Members serve 2-year terms (but can be dismissed at any time) and are eligible for re-appointment. The Council aims to include non-government workers who are members of a variety of fields and disciplines: science and medicine, law and government, philosophy and theology, and other areas of the humanities and social sciences. These members are not paid a salary, but they are compensated and all expenses they incur in serving on the Council are paid (i.e., travel to Council meetings). The funding and administrative support for this committee comes from the Department of Health and Human Services. Issues specifically within their purview are the moral significance of developments in biomedical technology and the ethical and policy questions related to these developments.

The President's Council is exactly that: it is the President who appoints its members. All of the its members were appointed by President George W. Bush, who is well known to have strong religious convictions against stem cell research as well as reproductive cloning. President Bush is thus likely to listen to their analysis (Bush 2001).

However, the President's Council is unlikely to represent all viewpoints. The current members are predominantly philosophers, physicians, and lawyers. Although advising on matters of genetics and embryology, as of June 2005 it has neither a geneticist nor an embryologist. Elizabeth Blackburn, a professor of cell biology, was abruptly dismissed from the Council in 2004 when she disagreed with the findings of the Council's position on stem cells (Blackburn 2004; see UCS 2004).

The council has been a conservative force and has been criticized for prioritizing religion over scientific knowledge and medical research (see Brainard and Blumenstyk 2004; Brainard and Smallwood 2004). The current chair of the Council is Leon Kass, a professor of philosophy and ethics at the University of Chicago and the author of *Life, Liberty, and the Defense of Dignity: The Challenge for Bioethics*. Kass believes that science can and does threaten the human condition, "both by undermining human self-esteem and by generating tools that might be misused, particularly by genetically reshaping the human mind or body" (quoted in Wade 2002). He publicly endorsed the ban on human cloning and the cloning of embryos for research when the bill was in the House of Representatives in July, 2001. Furthermore, he called the ban a way to "seize the initiative and to gain some control of the bio-technical project" (Kass 2002). One of President Bush's most recent appointment to the Bioethics Commission, Diana Schaub, is thought to be even more conservative than the President himself. She favors the outlawing of all forms of stem cell research and gene therapy, likening them to slavery (Holden 2004).

In October of 2003 the Genetics Policy Institute launched an initiative asking the World Court of the United Nations to rule on the question, "Does human reproductive cloning constitute a crime against humanity?" The Institute's argument was that human reproductive cloning is already illegal in the eyes of world law and needs to be labeled as such. At the time of the GPI proposal, the United Nations was also considering a treaty proposed by Honduras that would ban all human cloning, whether for therapeutic or reproductive purposes, as well as a less sweeping proposal that would ban only reproductive cloning. The Honduran measure was supported by the United States and 43 other countries. The ban on reproductive cloning only was supported by, among others, the United Kingdom, Japan, China, Germany, and France (as well as most of the world's scientific institutions).

After long debate in 2003, the United Nations adopted a compromise proposed by Iran that called for a two-year moratorium on cloning, to be reconsidered in 2005. Finally, in March 2005, the United Nations voted 84 to 34 (with 37 countries abstaining) in favor of a non-binding resolution that banned all forms of human cloning, both therapeutic and reproductive. The vote was carried largely by the United States and the predominantly Catholic countries (including those of Latin America). Most European nations voted against the resolution and said they would ignore it. Islamic nations were largely represented in the group that abstained from voting. An editorial in *Nature Cell Biology* portrayed the United Nations vote as a "missed opportunity to ban reproductive cloning" (Nature 2005).

Although the United States supported the United Nations measure to ban all cloning on a worldwide basis, the country has yet to pass any national laws on the issue (Grady 2004). Like the public, American politicians are almost uniformly against reproductive cloning, while they are split on whether or not to allow therapeutic cloning.

One of the major arguments against the simultaneous legalization of therapeutic cloning and banning of reproductive cloning is a "slippery slope" argument that says once therapeutic cloning is allowed, reproductive cloning would inevitably occur because the techniques would be available (Grady 2004). Others, like Daniel Perry, president of the Coalition for the Advancement of Medical Research, state that there is a "clear, bright line" between reproductive cloning and therapeutic cloning: the implantation of an embryo into a woman's uterus (Grady 2004). If this step is banned, says Perry, therapeutic cloning can be safely protected while reproductive cloning is avoided.

In a report commissioned for former President Clinton's National Bioethics Advisory Commission, Robert Cook-Deegan of the Institute of Medicine and National Academy of Science suggested that cloning be regulated but not banned. He put forth the argument that because many scientists may not agree with the reasons behind legislation that bans cloning, they would be liable to circumvent the laws and proceed without regulation (Cook-Deegan 1997).

Others also fear that banning reproductive cloning could lead to a dangerous black market for the technique and that the United States would be best advised to regulate rather than ban the technology. As Panayiotis Zavos, a fertility physician and outspoken proponent of human reproductive cloning, puts it, "the genie is out of the bottle." He argues that any reasonably high-tech in vitro fertilization clinic could perform a human cloning procedure today, and that legislation regulating reproductive cloning is the only way to keep it from becoming an unregulated, black-market practice (Zavos 2001). At this point, however, opposition to reproductive cloning is so strong, there has been very little discussion as to what form the cloning regulations might take.

Currently, the only federal law on the books in the United States is an appropriations bill that prohibits any cloning research from receiving federal funding. This legislation has no effect on privately funded research. Three major bills regarding cloning have been proposed in the House of Representatives, but none has been approved in the Senate.

The first bill, the Weldon-Stupak bill (H.R.2505), was passed in July of 2001 and was introduced in the Senate as the Brownback-Landrieu bill (S.1899). This bill called for a 10-year prison sentence or a million-dollar fine for anyone who performs any type of human cloning, reproductive or therapeutic (U.S. Senate 2002). This bill is not popular with scientists who feel that banning therapeutic cloning would needlessly limit progress in the field of human development. Rather, these scientists put their support behind the second major cloning bill to enter the Senate: the Specter-Feinstein bill (S.2439), which proposes to ban reproductive cloning while allowing therapeutic cloning (AAAS Policy Brief 2003).

Interestingly, the Specter-Feinstein bill was co-sponsored by interests as diverse as the conservative Republican senator Orrin Hatch and the liberal Democratic senator Edward Kennedy. In addition to the Specter-Feinstein bill, Kennedy and most of the other liberal senators sponsoring S.2439 are also sponsoring S.1758, a bill that strongly resembles S.2439, but with one difference: in addition to allowing therapeutic cloning, S.1758 has explicit guidelines for the regulation of therapeutic cloning commerce (U.S. Senate 2001). This distinction is critical, since it would be important to regulate therapeutic cloning as soon as such a law was passed. Senate Bill 1758 is widely supported by scientific societies (Letter in Support 2002), although it will need significant reworking to effectively allow therapeutic cloning while still blocking reproductive cloning. These bills leave enforcement and regulation up to other government departments, without many guidelines, suggesting that whether or not the laws were actually enforced might depend on the administration in power at the time.

In Conclusion

Reproductive cloning, which results in the birth of a new organism that is genetically identical to a single parent, is separate from therapeutic cloning, which seeks to create stem cell lines that can produce genetically compatible tissues and organs for those who need them. The dangers and low success rates so far experienced in mammalian reproductive cloning labs indicate that human reproductive cloning will not become widespread any time soon, but the first steps have certainly been taken. Over 30 nations have banned human reproductive cloning. The issue of whether to allow all cloning research or to ban it completely, or to allow therapeutic but not reproductive cloning research, is currently being debated in the United States.

UNIT 5

Should We Use Stem Cells To Repair the Body?

Between the fifth and tenth days the lump of stem cells differentiates
into the overall building plan of the embryo and its organs.
It is a bit like a lump of iron turning into the space shuttle.
In fact it is the profoundest wonder we can still imagine and accept,
and at the same time so usual that we have to force ourselves to
wonder about the wondrousness of this wonder.

MIROSLAV HOLUB (1990)

CHAPTER 9

BIOETHICS AND THE NEW EMBRYOLOGY

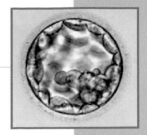

Regenerating Deficient Organs through Stem Cells

With few exceptions, every cell nucleus in the body contains the complete genome established at the time of fertilization. In molecular terms, the DNA of all differentiated cells within an organism is identical. This is vividly demonstrated by cloning. As we described in Chapter 7, complete and functional adult mammals have been generated from the nuclei of differentiated adult cells transplanted into enucleated oocytes. Thus, the unused genes in differentiated cells are neither destroyed nor mutated, and they retain the potential to become active and create an entire new individual. But if the DNA in all cells is identical, how do the cells become different? How is the inherited genome differentially expressed in different cells during the course of development?

The complete answer to this question is complex and occupies the research time of many developmental biologists. However, we do know that for vertebrates, the major pathway in the formation of the different organs and tissues involves stem cells. **Stem cells** are cells that have two important properties:

1. They have the capacity to divide for indefinite periods of time.

2. They have the ability at each cell division to give rise both to a similar stem cell as well as a more specialized cell type. That is, in addition to generating a more specialized type of cell, stem cells also generate more stem cells (Figure 9.1). This is crucial, because it means the population of stem cells is relatively constant so that more specialized cells can continually be made.

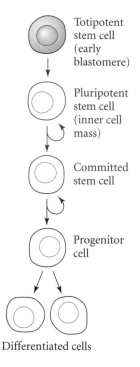

Totipotent stem cell (early blastomere)

Pluripotent stem cell (inner cell mass)

Committed stem cell

Progenitor cell

Differentiated cells

FIGURE 9.1   The stem cell concept. A generalized view of the cascade from totipotent to pluripotent to committed stem cell, through progenitor cell, to differentiated cell. When stem cells divide they form both a more committed cell and another stem cell (indicated by the circular arrow).

Stem cells are common in the embryo. Stem cells are also found among those adult cell types that are constantly being replaced. For example, stem cells allow our blood cells to be replaced continually, with about 2 million new red blood cells being made *each second* to replenish the same number that are destroyed. Similarly, we have stem cells for the epidermis of our skin. (Each of us sheds about a gram of skin cells every day, most of it becoming "house dust.") We have stem cells in our hair follicles so that we can replace the hair shaft that is pulled out. Human males have germinal stem cells that continually make new sperm, although females do not have germinal stem cells (and thus have a limited number of eggs that are ovulated each month after puberty; see Chapter 3). Recent evidence shows that we also retain some neural stem cells in our brains and some muscle stem cells lining our muscles.

The Biology of Stem Cells

There are different types of stem cells in the mammalian embryo. Each cell of an early embryo is capable of becoming every type of cell in the body, as well as that of the trophoblast (the fetal placenta). These cells—such as the first eight cells of the human embryo—are called **totipotent stem cells** (see Chapters 1 and 7). Once the interior cells have formed the inner cell mass and the external cells have become the trophoblast (see Figure 1.8), the cells of the inner mass have the ability to produce all the cells of the embryo (but not those of the trophoblast). These inner mass cells are now called **pluripotent stem cells**. It is these pluipotent cells that, when cultured in the laboratory, are called **embryonic stem cells**, or **ES cells**.

Pluripotent stem cells interact with one another and with the trophoblast. They produce **lineage-restricted stem cells**. For instance, a mesodermal stem cell can no longer form ectodermal (epidermal or neural) cell types nor endodermal (gut, lung) cell types. Rather, they become restricted to forming the mesodermal organs (heart, blood, immune cells, bone, muscles, dermis, kidney, and gonads). Then, through the interactions between cells, the progeny of a mesodermal stem cell might be restricted to becoming blood cells or blood vessels

Figure 9.2 outlines the restriction of a pluripotent stem cell lineage into red blood cells. The totipotent stem cell gives rise to a lineage-restricted

FIGURE 9.2   Stem cells are critical in cell differentiation in humans. At each stage, there is a restriction in potency. Here, the differentiation of a red blood cell is depicted as proceeding through progressively more limited stem cells, until a progenitor cell is formed that has *only* red blood cells as its progeny.

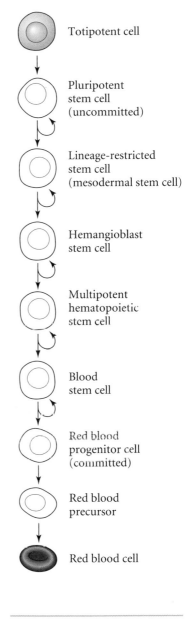

Totipotent cell

Pluripotent stem cell (uncommitted)

Lineage-restricted stem cell (mesodermal stem cell)

Hemangioblast stem cell

Multipotent hematopoietic stem cell

Blood stem cell

Red blood progenitor cell (committed)

Red blood precursor

Red blood cell

stem cell that will produce mesoderm. The lineage-restricted mesodermal stem cell thus can become a somewhat more restricted hemangioblast (from the Greek *haema*, blood + *angion*, vessel + *blastos*, multiplying) stem cell. Hemangioblast stem cells interact with one another such that some of these cells give rise to the blood vessels; other hemangioblast stem cells produce a cell type that only gives rise to blood cells: the hematopoietic stem cell (again from the Greek: *haema*, blood + *poiein*, to create). These cells interact with other cells in its vicinity to generate a stem cell that only makes red blood cells. The hematopoietic cell is a **committed stem cell**. It is a stem cell (since it can divide to produce both a more differentiated type of cell as well as another stem cell of the same type), but the type of cell will always be the same. In this case, the red blood stem cell will divide to produce another red blood stem cell as well as a red blood progenitor cell. **Progenitor cells** are capable of cell division, but they are not stem cells. Each of the two daughter cells of a progenitor cell will eventually differentiate into red blood cells.

Therefore, cell specification involves the successive restriction of what a cell can become, and these decisions are made by **cell-cell interactions**. The way that the stem cells are told to divide and to produce specific cell types is through the communication between cells. In other words, cells tell each other what to become.

Paracrine Factors and Cell-Cell Interactions

The "conversations" between cells are mediated by a family of molecules called **growth factors** or **paracrine factors**. Paracrine factors are usually proteins, and they are secreted by one cell type and are received by neighboring cells. In a sense, they are like hormones, which also signal the cells they interact with to change. But while hormones (which are sometimes called "endocrine factors") travel through the bloodstream to reach their target cells, paracrine factors act locally. In some instances, paracrine factors decay so rapidly that the

Growth factors will undoubtedly be the magic potions of the twenty-first century.

JONATHAN SLACK (1999)

cell producing the factor must be touching the receiving cell if the signal is going to work.

There are many different paracrine factors. Moreover, in order to respond to a paracrine factor, the cell receiving the paracrine signal must have a receptor for it.* Not all cells will have such receptors. In this way, certain cells will be capable of responding to a paracrine factor to become a new type of cell. Other nearby cells, lacking the specific receptor, will be incapable of responding to it. This type of development, where the cells are instructed by their neighbors, is called **induction**. It is the most common way that cell fates are determined in vertebrate embryos.

An example: VEGF and blood vessel formation

Earlier we mentioned hemangioblast stem cells that become restricted to form the blood vessels. These presumptive vessel cells have a receptor that will bind a paracrine factor called vessel endothelial growth factor (VEGF). VEGF is produced by certain cells in the area of the mesodermal stem cells. Hemangioblast cells that have the receptor and come in contact with the VEGF protein will differentiate into endothelial cells (the "lining" of the blood vessels); the VEGF paracrine factor also stimulates the committed vessel cells to multiply. (Should an embryo have a mutation in its gene for VEGF factor or in the VEGF receptor gene, it will not be able to make blood vessels and will die within the uterus.)

But the signaling is not yet finished. Once the newly formed endothelial cells have generated the lining of the blood vessel, they in turn secrete a different paracrine factor called angiopoietin. Angiopoietin instructs the cells around the blood vessel to become the smooth muscle cells of the blood vessels—cells that are necessary to move blood through the vessels.

Mechanisms of paracrine induction: Transcription factors

The basic story of how paracrine signals work is very simple. Receptors have three domains. The first domain is inside the cell. The second domain spans the cell membrane, and the third domain extends outside the cell (Figure 9.3). When a paracrine factor binds to this outside "extracellular" domain of the receptor, it causes the inside domain of the receptor to change its shape. When the internal domain has a new shape, it also has a new activity. It can now act as an enzyme to promote certain chemical reactions. These reactions usually add a phosphate group to another, and very special, set of proteins called **transcription factors**. Once a transcription factor has a phosphate group on it (has been "phosphorylated," in biochemi-

---

*This is also true of the signals from hormones. See the case of androgen insensitivity syndrome described in Chapter 5 (page 90).

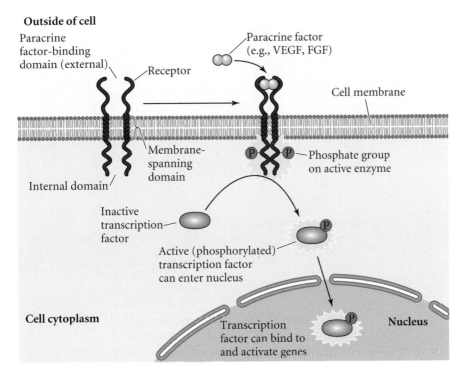

FIGURE 9.3   The principle of paracrine factor signaling and transcription factors. A receptor protein spans the cell membrane. The receptor binds a paracrine factor outside the cell, causing changes to the structure of the protein domain within the cell and triggering enzymatic chemical activity by the protein. This new enzyme activity adds phosphate groups to itself and to the transcription factor, activating the transcription factor and allowing it to enter the cell nucleus. Once within the nucleus, the transcription factor can stimulate (or repress) the activity of a given gene ("turning genes on or off").

cal terms), it becomes active and can enter the cell nucleus. Once inside the nucleus, the transcription factor can bind to an area of the gene known as the **enhancer**, thus turning the gene on (or off). Enhancers will be discussed in more detail in Chapter 11, but this brief explanation shows how a paracrine factor from outside the cell can determine which genes are going to be expressed inside a given cell.

Reciprocal induction

In many cases, some of the genes activated by a transcription factor are genes that make other paracrine factors. These new paracrine factors can then bind to the cells that originally produced the first paracrine factors; in this way, adjacent cells can

All that you touch
You Change
All that you Change
Changes you.

OCTAVIA BUTLER, *THE PARABLE OF THE TALENTS* (1998)

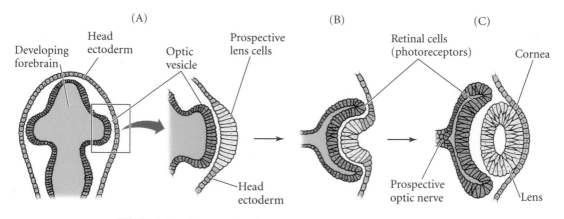

FIGURE 9.4   Reciprocal induction of the eye. (A) Paracrine factors induce a portion of the head ectoderm to become lens tissue. (B) Prospective lens tissue produces different paracrine factors that induce retinal cells. (C) Factors from the retina induce differentiated lens tissue, which in turn causes the retina to fold and help induce the cornea in front of the lens.

interact with one another to create organs. Thus, the cell that produced the VEGF that induced the endothelial vessel cells can now be induced by the angiopoietin secreted by these same endothelial cells to become the smooth muscles that line the blood vessels. This **reciprocal induction** is critical for coordinating cell differentiation and organ formation.

For instance, during development in vertebrate animals, certain neural cells bulge from the brain and form an optic vesicle. The cells of the optic vesicle contact the ectoderm (outer layer) of the head and put forth paracrine factors to induce those head ectoderm cells they touch to become the early lens tissue of the eye instead of skin epidermis (Figure 9.4A). The newly induced presumptive lens cells then fold inward into the head and in turn produce paracrine factors that signal those neural cells that induced them to become the light-perceiving cells of the eye's retina (Figure 9.4B). These new retinal cells then produce factors telling the initial lens cells to keep differentiating into lens tissue. The differentiating lens tissue then produces factors that induce the head ectoderm cells above them to become the cornea (Figure 9.4C). Thus, an organ such as the eye is formed by a cascade of reciprocal inductive interactions mediated by paracrine factors.*

---

*Creationists have put forth the argument that an organ as complex as the vertebrate eye could not possibly have been generated by natural selection, but must have been created *de novo*, intact and in its functional form. In point of fact, the processes of reciprocal induction seen in development throughout the animal kingdom make possible the exact positioning of the lens, cornea, and retina, as well as the musculature and the blood vessels maintaining them. These processes are most certainly capable of producing complex structures that evolve through natural selection.

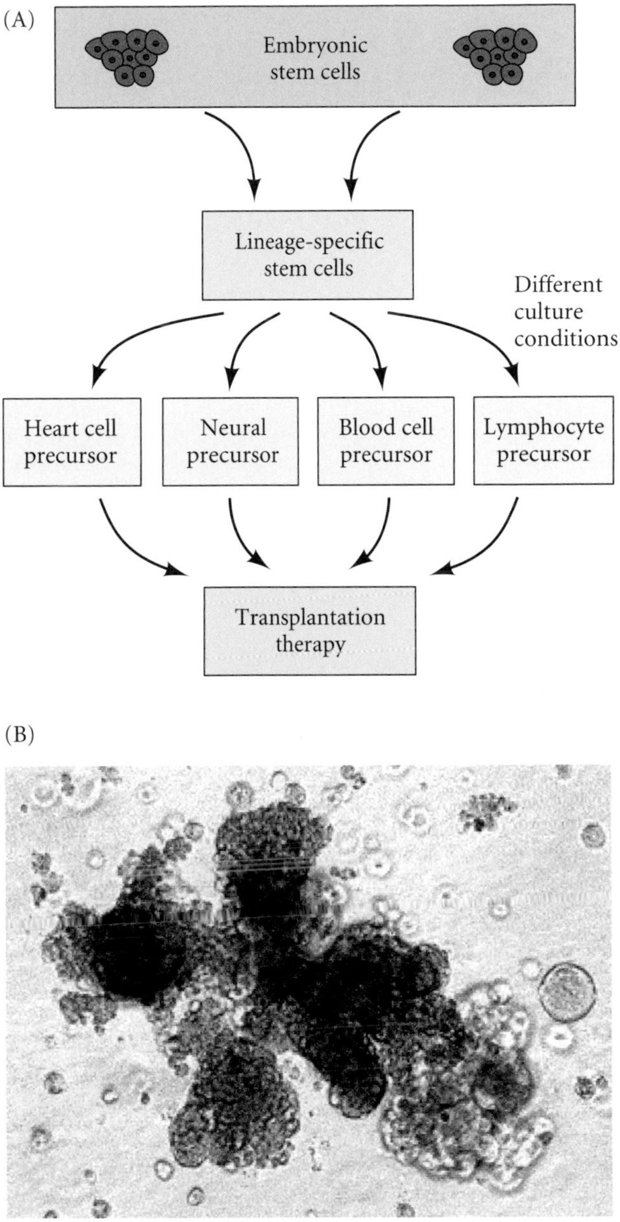

FIGURE 9.5 Embryonic stem cell therapeutics. (A) Human embryonic stem (ES) cell cells can differentiate into lineage-specific stem cells that could then be transplanted into the host needing replacement cells. (B) Blood cells developing from human embryonic stem cells cultured on mouse bone marrow. (Photograph courtesy of the University of Wisconsin.)

This brief overview of paracrine signaling reveals one of the reasons embryonic stem cells are important: if scientists knew all of the appropriate paracrine factors in the pathway that creates a particular cell type, they could take an embryonic stem cell, apply the factors, and stimulate the stem cell to make the desired cell type (blood cells or neural tissue, for example; Figure 9.5). These new cells could be transplanted to replace diseased or missing tissues.

Embryonic Stem Cells and Therapeutic Cloning

The potential importance of embryonic stem cells for medicine is enormous. A few of the possible scenarios include using human ES cells to produce new neurons for patients with degenerative brain disorders (such as Alzheimer's or Parkinson's disease) or spinal cord injuries; to produce new pancreatic cells for those with diabetes; and to produce new blood cells for people with anemias. Persons with deteriorating hearts might be able to replace the damaged tissue with genetically identical heart cells, and those suffering from immune deficiencies might be able to replenish their failing immune systems.

Embryonic stem cells are pluripotent and can be cultured for long periods of time in an undifferentiated state. Human embryonic stem cells can be obtained from two major sources. First, they can be derived from the inner cell masses of human embryos (Figure 9.6A). The source of these cells is typically the embryos left over after in vitro fertilization procedures, since these procedures (as described in Chapter 3) generate many more embryos than are actually transplanted. Second, embryonic stem cells can be generated from germ cells derived from fetuses that have miscarried (Figure 9.6B). In

FIGURE 9.6   The major ways that have been proposed to obtain human pluripotent stem cells. Functional stem cells have already been derived using paths A and B. Path C remains experimental. (A) The inner mass cells of the blastocyst-stage embryo are cultured and become pluripotent embryonic stem cells. (B) The primordial germ cells of fetuses are harvested and become pluripotent embryonic stem cells. (C) Adult stem cells can be obtained from certain organs or tissues. It is possible that some of these cells can be cultured so as to become pluripotent. (D) Therapeutic cloning, in which the nucleus of a somatic cell is transferred into an enucleated oocyte. The activated oocyte gives rise to a blastocyst whose inner cell mass is harvested and cultured to become pluripotent embryonic stem cells. (After NIH, 2000.)

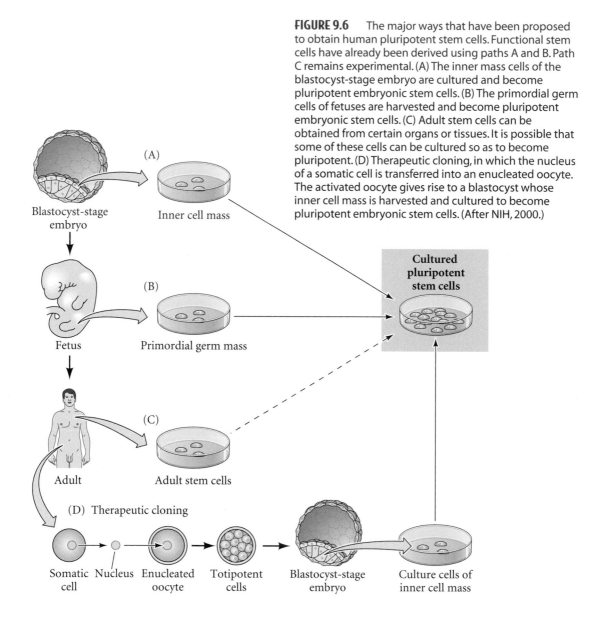

both these instances, the embryonic stem cells are pluripotent, since they are able to differentiate in culture to form more restricted stem cells.

Stem cell therapy has already worked in mice. Mouse ES cells have been cultured in conditions causing them to form lineage-specific stem cells capable of producing insulin-secreting pancreatic cells, muscle cells, glial cells, and neural cells (Brüstle et al. 1999; McDonald et al. 1999). For instance, when mouse ES cells were placed in a dish containing two paracrine factors—platelet-derived growth factor and basic fibroblast growth factor—the embryonic stem cells divided into *glial* stem cells (Figure 9.7A). (Glial cells support the nervous system, maintain the neurons, and may play an important role in memory storage.) If the same ES cells were cultured in a medium containing a different paracrine factor—retinoic acid—they became *neural* stem cells (Figure 9.7B). Most importantly, these glial and neural stem cells were functional. When placed into the brains of diseased mice, they were able to restore glial and neural functions. Indeed, neurons derived from mouse embryonic stem cells have been shown to significantly reduce the symptoms of a Parkinson-like disease in mice (Bjorkland et al. 2002; Kim et al. 2002).

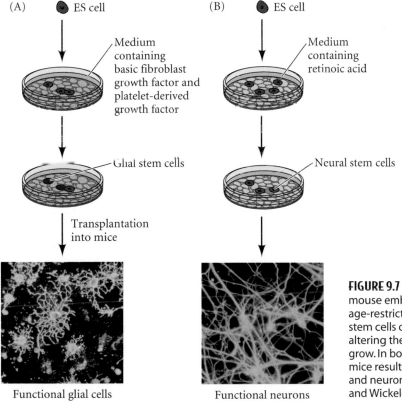

FIGURE 9.7 The differentiation of mouse embryonic stem cells into lineage-restricted glial (A) and neural (B) stem cells can be accomplished by altering the media in which the ES cells grow. In both cases, transplantation into mice resulted in functional cells (glia and neurons). (From Brüstle et al. 1999 and Wickelgren 1999.)

Although human embryonic stem cells differ in some ways from their mouse counterparts in their growth requirements, in most ways they are very much alike. Like mouse ES cells, human ES cells can be directed down specific developmental paths. For example, researchers have been able to direct human embryonic stem cells to become blood-forming stem cells (Kaufman et al. 2001). These hematopoietic stem cells could further differentiate into numerous types of blood cells (see Figure 9.5C). In other experiments, cells secreting insulin have been generated from human embryonic stem cells (Assady et al. 2001).

One big difference between laboratory mice and humans is that lab mice are inbred and genetically identical. Humans, obviously, are not. This means that as human ES cells differentiate, they express significant amounts of the certain proteins that can cause immune rejection. These proteins, found on the membrane surfaces of the body's nucleated cells, are the products of the **major histocompatibility complex** (**MHC**) genes of the immune system. One of the major attributes of MHC genes is their ability to generate an enormous variety of protein combinations; this ability makes them virtually unique in each individual. (MHC genes are thus the basis for much of DNA forensics—"DNA fingerprinting"—and for paternity testing.) Although this variability is intrinsic to the proteins' ability to help the body fight off bacteria, viruses, and other foreign agents, it also means that these proteins attack, or reject, any substance they perceive as foreign to the body, including transplanted tissues and organs that are not genetically identical to the recipient.

To get around the problem of rejection, human ES cells either have to be modified so that they no longer express the MHC proteins that cause graft rejection, or they have to be made genetically identical to the patient. This genetic alteration can be done through somatic cell nuclear transplantation (i.e., cloning). By putting a nucleus from the patient's own somatic cells into an activated enucleated oocyte, that oocyte could make the blastocyst from which embryonic stem cells are taken. In this way, the stem cells are genetically identical to the person who will receive their cellular progeny (see Figure 9.6D). This use of somatic cell nuclear transplantation has been called **therapeutic cloning**. Unlike reproductive cloning, the goal of therapeutic cloning is not to create a new individual (Table 9.1). Rather, the goal is to get stem cells from an early embryo that is genetically identical to the person needing treatment.

Recent experiments in mice have shown that paracrine factors can direct the differentiation of embryonic stem cells into neurons capable of replacing those destroyed by Parkinson's disease. Parkinson's disease is one of the most prevalent neural degenerative diseases, afflicting about a million Americans. In this disease, dopamine-producing neurons of the substantia nigra (a cluster of cells in the base of the brain) are destroyed. The lack of functional dopamine leads to muscle tremors, difficulty in initiating voluntary movements, and problems in cognition. Injection of the drug L-DOPA

TABLE 9.1 Materials and techniques of stem cell research and cloning (SCNT)[a]

Technique	Purpose	Starting material	End product
Adult (or fetal) stem cell research	To obtain undifferentiated stem cells for research and therapy	Isolated stem cells from adult or fetal tissue	Cells produced in culture to repair diseased or injured tissue
Embryonic stem (ES) cell research	To obtain undifferentiated stem cells for research and therapy	Stem cells from a blastocyst-stage embryo	Cells produced in culture to repair diseased or injured tissue
Therapeutic cloning ("nuclear transplantation")	To obtain undifferentiated stem cells that are genetically matched to the recipient for therapy and tissue regeneration	Stem cells from a blastocyst-stage embryo produced from an enucleated egg supplied with nuclear material from the recipient's own somatic cells	Cells produced in culture to repair diseased or injured tissue
Reproductive cloning (embryos produced using SCNT; see Chapter 7)	To produce an embryo for implantation in womb, leading to birth of a child	Enucleated egg supplied with material from a donor somatic cell	Embryo (and eventual offspring) genetically identical to donor of nuclear material

Source: National Institutes of Health 2001: *Stem Cells: Scientific Progress and Future Research Directions*

[a] SCNT is the abbreviation for "somatic cell nuclear transfer," the technical term for the technique used to clone cells or organisms. The distinction between therapeutic and reproductive cloning is particularly important.

(which the body metabolizes into dopamine) relieves the symptoms temporarily, but ʟ-DOPA loses its effect with prolonged use, and it sometimes has adverse side effects.

One possibility for curing Parkinson's disease would be to take the nuclei of patient's own somatic cells, transplant the nuclei into oocytes, let the oocytes develop into a blastocyst with an inner cell mass, and produce embryonic stem cells from the inner cell mass. These embryonic stem cells would then be cultured in a mixture of paracrine factors that would induce them to become the cells committed to producing dopaminergic (dopamine-synthesizing) neurons. These committed stem cells would then be transplanted into the appropriate region of the brain, where they could restore proper brain function. Laboratory studies have shown already that this is possible in mice (Barberi et al. 2003). Barberi and his colleagues induced cloned embryonic stem cells to become neural stem cells by growing them on mesoderm cells and providing them with a particular paracrine factor, fibroblast growth factor-2 (FGF2; Figure 9.8). The neural stem cells were then induced to become ventral (lower brain) neural cells

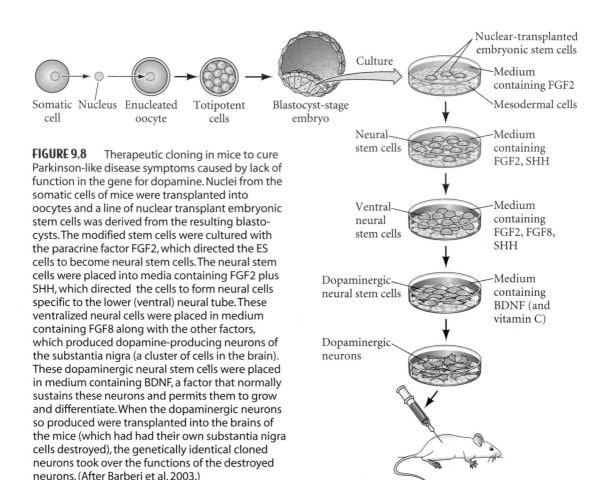

FIGURE 9.8   Therapeutic cloning in mice to cure Parkinson-like disease symptoms caused by lack of function in the gene for dopamine. Nuclei from the somatic cells of mice were transplanted into oocytes and a line of nuclear transplant embryonic stem cells was derived from the resulting blastocysts. The modified stem cells were cultured with the paracrine factor FGF2, which directed the ES cells to become neural stem cells. The neural stem cells were placed into media containing FGF2 plus SHH, which directed the cells to form neural cells specific to the lower (ventral) neural tube. These ventralized neural cells were placed in medium containing FGF8 along with the other factors, which produced dopamine-producing neurons of the substantia nigra (a cluster of cells in the brain). These dopaminergic neural stem cells were placed in medium containing BDNF, a factor that normally sustains these neurons and permits them to grow and differentiate. When the dopaminergic neurons so produced were transplanted into the brains of the mice (which had had their own substantia nigra cells destroyed), the genetically identical cloned neurons took over the functions of the destroyed neurons. (After Barberi et al. 2003.)

"Parkinsonian" mouse (cannot walk) injected with genetically identical dopaminergic neurons

Walking mouse

by the addition of the paracrine factor known as sonic hedgehog (SHH). Further exposure of these cells to paracrine factors FGF2 and to FGF8, followed by exposure to another paracrine factor, brain-derived neurotrophic factor (BDNF) produced cells that had all the characteristics of dopaminergic neurons. Moreover, when these cells were injected into mice that had

had their dopaminergic neurons destroyed, the ES-derived neurons were able to restore normal function to the mice.

A similar study has shown that embryonic stem cells derived from monkey embryos can cure a Parkinson-like disease induced in adult monkeys whose dopaminergic neurons had been chemically destroyed (Takagi et al. 2005). Thus, nuclear transplanted embryonic stem cells appear to be able to provide a reusable and readily available source of genetically identical dopaminergic neurons. This availability may be very important should the treatment ever come into use: because the cause of the destruction of dopaminergic neurons in Parkinson's disease has not been established, it is possible that a patient might have to have several such replacements of dopaminergic neurons. The Hwang laboratory in South Korea has shown that such therapeutic cloning procedures could be made efficient enough for human medicine. Hwang and his colleagues recently made 11 stem cell lines that were genetically identical to patients suffering from spinal cord injuries and insulin-deficient (Type 1) diabetes (Hwang et al. 2005). The researchers transferred nuclei from patients' somatic cells into enucleated oocytes and grew these embryos to the blastocyst stage. Stem cell lines were derived from the inner cell masses of these blastocysts. The hope is that these cultured stem cell lines can form precursor cells that will grow back neurons or pancreatic cells and cure the respective diseases of these patients.

Interestingly, the methylation problems that plague cloned animals (see pages 122–123) do not appear to be a problem for embryonic stem cells that are derived from nuclear transfer. It appears that the nuclei within the small population of ES cells that survive in culture have had their methylation patterns erased, thus enabling them to re-differentiate (Jaenisch 2004).

Adult Stem Cells

So far, we have focused on the potential of embryonic stem cells. However, certain relatively undifferentiated cell lineages remain in our bodies into maturity. Adult stem cells are cells that are held in reserve until they are needed for tissue repair or renewal. Most tissues are capable of some degree of renewal to replace damaged or worn out cells. Tissues such as the epidermis (the outer layer of the skin), the lining of the gut, and blood all have a high turnover rate and require a constant source of stem cells for rapid renewal. Even some tissues with slower rates of renewal, such as the liver and bones, also rely on stem cells.

Bone marrow stem cells

Perhaps the most remarkable source of stem cells found in the adult is in the bone marrow: the hematopoietic (blood-forming) stem cells of the bone marrow renew all the cells of the blood and immune systems (see Figure

9.2). One type of stem cell can give rise to the white blood cells, the red blood cells, the platelets, and the lymphocytes. This stem cell type is rare, perhaps found less than once out of every 15,000 bone marrow cells (Spangrude et al. 1988); but injection of a very few (or even one) of these cells into anemic mice will cure them of anemia (Osawa et al. 1996). Similarly, in humans, people suffering from red blood cell deficiency or leukemias can have their blood system restored by the injection of hematopoietic stem cells (Parsons 2004). Indeed, the paradigm for adult stem cell therapy comes from bone marrow transplantation, wherein about 40,000 such transplants are now performed each year.

Recent studies have indicated that the bone marrow may contain even more surprising stem cells. When biopsies were taken from individuals who had received hematopoietic stem cell transplants, researchers found indications that some liver and skin cells in these patients were also derived from the bone marrow donor cells (Korbling et al. 2002). Certain experiments with mice indicated that transplanted bone marrow stem cells could become muscle stem cells and even mature muscle cells (LaBarge and Blau 2002); other experiments found donor cells contributing to the intestinal tissue of mice.

These findings could be interpreted to mean that, in addition to hematopoietic stem cells, there is a distinct mesenchymal stem cell found in bone marrow that can give rise to cartilage, bone, fat cells, and connective tissue, but not blood cells (Pittenger et al. 1999). However, some laboratories doubt the existence of such a pluripotent mesenchymal stem cell in the bone marrow, and research is ongoing. Although these results are preliminary, adult bone marrow remains a promising source of stem cells for regenerative medicine.

Even adult stem cells with limited potency can still be very useful. Carvey and colleagues (2001) have shown that when neural stem cells from the midbrain of adult rats are cultured in a mixture of particular paracrine factors, they differentiate into dopaminergic neurons that can cure the rat version of Parkinson's disease. Techniques to specifically amplify the multipotent stem cells already present in various organs would make such therapy applicable in humans.

Adult versus embryonic stem cells

There are advantages and disadvantages to both the embryonic stem cells and the adult stem cells. Scientists who advocate research on embryonic stem cells point out that these cells can be obtained in relatively large quantities. Moreover, they can point to successes in animal experiments in which ES cells have differentiated under defined conditions and have cured the animal's disease.

Those critical of embryonic stem cells point out that obtaining them involves the ethically ambiguous area of destroying very young embryos

Stem Cells from the Umbilical Cord

In addition to pluripotent embryonic stem cells, multipotent adult stem cells, and possibly pluripotent adult stem cells, there may be at least two other types of stem cells. First, it appears that the fetus sends both lineage-specific and pluripotent stem cells into the mother. Studies have shown that these pluripotent stem cells persist in the maternal circulation even after pregnancy, and that when harvested, they are capable of differentiating into liver cells, intestinal cells, thyroid cells, and uterine cells (Khosrotehrani and Bianchi 2005).

Blood from the umbilical cord has also been found to contain pluripotent stem cells (Edwards 2004; Kogler et al. 2004). Some physicians now propose to prospective parents that they can "bank" their child's umbilical stem cells, freezing and storing them so that the child's own stem cells are available to be used if a transplant of any kind is needed in the future. In fact, the *Korean Times* reported in 2004 that human umbilical cord stem cells had differentiated into neural stem cells that, when transplanted into a paraplegic woman, allowed her to walk (Tae-Gyu 2004).

and that to be useful in treatment the ES cells would have to be made genetically identical to the patient, probably by somatic cell nuclear transfer (cloning). There also appears to be a risk that, if the transplant dosage is not exact, transplanted ES cells can form tumors in the recipient. There is also some concern that the ES cells are grown on a medium of "feeder" cells, which might transmit viruses to the ES cells. These latter three problems may just be technical matters that new research will refine (see Solter and Gearhart 1999; Shamblott et al. 2001).

Those scientists advocating using adult stem cells point to bone marrow transfusion as the paradigm for human stem cell therapy. They argue that obtaining adult stem cells does not involve the production and subsequent destruction of new embryos. Furthermore, if the stem cells could be obtained directly from the patient being treated, there would certainly be no problem with genetic identity between the patient and the new tissue.

The critics of using adult stem cells point out that such cells are relatively rare; that they are not easily grown outside the body; and that certain organs do not have adult stem cells. This deficiency appears to be the case with the pancreas, whose beta cells are responsible for making insulin. There was some hope that adult pancreatic stem cells could be used to cure insulin-dependent diabetes. However, pancreatic tissue, including the beta cells, does not seem to be replaceable using adult stem cells (Dor et al. 2004).

The discovery of stem cells and paracrine factors has revolutionized our view of human development. It has also opened up the possibility for new medical treatments and increased life expectancy. In the past few years, experiments in different laboratories have been able to repair heart defects in animals using embryonic stem cells (Fraidenraich et al. 2004), bone marrow stem cells (Silva et al. 2004), and umbilical cord-derived stem cells (Hirata et al. 2005). Thus it is certainly premature to say that only one avenue of research is going to produce important medical results, much less to predict which avenue will be fruitful.

In fact, saying that scientists should be able to study *only* embryonic stem cells or *only* adult stem cells appears to be an emotional rather than a reasoned decision at this moment. Yet people are indeed saying such things, as we will evaluate in the next chapter as we look at some ethical issues raised by stem cell research.

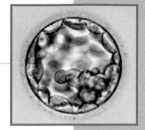

Ethical Dilemmas in Stem Cell Therapy

We are at an unusual crossroads in modern medicine, a point at which the medical dictum *primum non nocere* (most importantly, do no harm), appears to some to come into conflict with heretofore undreamed of possibilities to cure some of the most debilitating and widespread medical conditions in modern society. Certain cells of bastocyst-stage human embryos have been shown to be capable of almost unlimited growth in culture, and their ability to differentiate into any tissue of the body bodes well for their being a source of regenerative tissue for deteriorating organs.

Embryonic stem cell research and therapeutic cloning hold out the promise of medical treatments that could alleviate or even eliminate conditions including Parkinson's and Alzheimer's diseases, multiple sclerosis, diabetes, certain heart conditions, and traumatic spinal cord injury (National Institutes of Health 2000). Research universities, biotechnology companies, and medical institutions generally are anxious to push the field forward; the public at large has been more cautious as they slowly become aware of both the positive implications and potential hazards of the work. Does the destruction of a human embryo at the very earliest stages of development constitute harm that is morally unacceptable when weighed against the potentially monumental gains in the war against human suffering?

Embryonic stem (ES) cell research is only one sector of a larger field of stem cell research that is progressing rapidly, as described in Chapter 9. There are several ethical issues presented by the possibilities of stem cell therapy. One of the most critical problems involves the sources of stem cells. Should we pursue embryonic stem cell research, or limit our research to adult stem cells? Another issue concerns the economics of egg donation. In a technolo-

[R]esearch involving human pluripotent stem cells … promises new treatments and possible cures for many debilitating diseases and injuries, including Parkinson's disease, diabetes, heart disease, multiple sclerosis, burns, and spinal cord injuries. The NIH believes the potential medical benefits of human pluripotent stem cell technology are compelling and worthy of pursuit in accordance with appropriate ethical standards.

NATIONAL INSTITUTES OF HEALTH (2000)

gy that uses hundreds of eggs, who will supply these eggs, how might the suppliers be paid, and do these donors have rights? A third important ethical issue concerns life extension. If stem cell therapy works, would it in fact be a good thing to allow people to extend their natural life span, living much longer with the help of repeated stem cell therapies?

Sources of Embryonic Stem Cells

Some who support human embryonic stem cell research argue that because embryonic stem cells are isolated from an early-stage embryo, prior to any differentiation into organs, and because once isolated these cells are incapable of giving rise to an embryo, the moral issue of whether the embryo has acquired "personhood" is avoided. Some religions argue against using human embryos based on the belief that "personhood" begins at fertilization (see Chapter 2). However, some scientists see no moral dilemma at all. In an editorial for the journal *Science*, Donald Kennedy of Stanford pointed out that "this well-advertised dilemma does not arise from a confrontation between science and ethical universals. Instead, the objections arise from a particular belief about what constitutes a human life: a belief held by certain religions and not by others" (Kennedy 2005).

Neither side argues that the human embryo is without value, and accompanying every argument is the dictum that the human embryo should always be afforded proper respect and never be commercialized or treated like a commodity. And even among supporters of human embryonic stem cell research, there is argument over what constitutes an acceptable source of embryos for this research. There are currently two identifiable embryo sources: embryos that are "left over" after in vitro fertilization (IVF) procedures, and embryos created solely for obtaining stem cells.

The numerous IVF clinics worldwide that help infertile couples conceive create more embryos than are ever implanted (see Chapter 4). The question arises, what should be done with the unused (supernumerary) embryos? These embryos are routinely frozen, and when the couple no longer has a use for them, they can either be donated to another hopeful parent or destroyed. To many, supernumerary embryos from IVF clinics constitute a suitable source of embryonic stem cells, since the intent behind their creation affords proper respect and does not risk objectifying the embryo as a source of material for commercial use.

Ethical review committees that have recommended this use of supernumerary IVF embryos advise specific guidelines to insure proper respect for the embryo (National Institutes of Health 2000; European Group on Ethics 2002). It is recommended, for example, that the couple donating the embryos not be informed of this choice until after they have decided to have the embryos discarded and that no money be given the couple donating the embryos so as to avoid possible coercion of the poor.

Some researchers argue that the unused embryos from IVF will not provide a broad enough range of genetic types for stem cell research or therapy. In this argument, the only way to ensure a broad range of genetic types is to allow the creation of embryos specifically for the isolation of human embryonic stem cells.

The embryonic stem cells used in therapeutic cloning must come from an embryo created solely as a source of these cells. As detailed in Chapter 9, therapeutic cloning inserts the nucleus from a differentiated somatic cell *taken from the patient* into a donor's egg cell that has had its nucleus (and thus its genes) removed. The fused egg is then stimulated to divide in culture. When the embryo had been cultured to the blastocyst stage (approximately 5 days; see Figure 1.4), the cells of the inner cell mass are removed and cultured to create a population of stem cells. These stem cells might then be induced to form the specific tissue needed for transplantation by adding exogenous inducers (see pages 147–149). In therapeutic cloning, tissues, not people, are the intended result. Research on the specific inducers and culture conditions needed to cause the differentiation of specific tissue types is still in its infancy but is advancing rapidly (see Wasserman and Keller 2003).

Those who argue in favor of therapeutic cloning point out that the embryo is not being created for implantation, besides which its viability as an embryo is already severely compromised because its nucleus was transferred from an already differentiated cell (see Lanza and Rosenthal 2004). One argument holds that, because babies are never intended to arise from these embryos, their destruction is not objectionable. Paul McHugh of the President's Council on Bioethics coined the term "clonote" to be distinct from the term zygote. He asserts that clonotes are "manufactured rather than begotten;" they are meant for research not reproduction, and it is therefore not immoral to kill them. Rudolph Jaenisch, a pioneer of cloning research, also upholds this position and states that in his opinion, "the destruction of a cloned embryo to make embryonic stem cells poses less ethical problems than the destruction of frozen embryos in the IVF clinic" (Hall 2004).

Those who argue against therapeutic cloning point out that knowledge gained in this field will advance the field of reproductive cloning, increasing the likelihood of a successful attempt at human reproductive cloning, which the vast majority of people consider to be unacceptable (see Chapter 8). For those who are against using human embryos for therapeutic use based on the belief that human life begins at the moment of conception, therapeutic cloning poses a confusing dilemma, since by not involving the joining of an egg and sperm, conception technically never takes place. To date, most ethics committees deem that therapeutic cloning still poses too great a risk and advise that research proceed using supernumerary IVF embryos to create human embryonic stem cell lines so that the safety and effectiveness of the procedures can be evaluated (National Bioethics Advisory Commission 1999; Commission of the European Communities 2003).

One question that can be asked is whether a cloned embryos would be a more acceptable source of stem cells if the genetic material were first modified in such a way that the embryo could only give rise to stem cells—that is, even if the embryo were to be implanted it would not develop into a fetus. Research on mice indicates that such a thing may be possible.

A recent proposal (2004) by William Hurlbut, a member of the President's Council on Bioethics, seeks to create human stem cell lines from human embryos that lack the *CDX2* gene. His suggestion is based on the fact that the mouse *Cdx2* gene* is essential for development of the trophoblast, and the embryo cannot develop beyond the blastocyst stage without it. However, *Cdx2*-deficient embryos can still generate embryonic stem cells from the inner cell mass (Chawengsaksophak et al. 2004). If humans in fact developed in the same way, then such cells would be able to be used for human embryonic stem cells, but would not be able to form a human if transplanted into a uterus.

However, biologist Douglas Melton and his colleagues have pointed out that we don't actually know the functions of the *CDX2* gene in humans, and to find out, we would have to experiment on human embryos. Moreover, they note that even in the mouse, the *Cdx2* gene has several functions, and it also is working in the mouse embryo itself. They argue that there is "no basis for concluding that the action of *CDX2* (or indeed any other gene) represents a transition point at which a human embryo acquires moral status" (Melton et al. 2004).

Another possible source for embryonic stem cells was discovered by a small biotech company in Massachusetts, Advanced Cell Technology (ACT). In 1998, ACT announced that they had fused human nuclei with enucleated cow oocytes and that these hybrid cells had divided in culture, making them a potential source of human embryonic stem cells (Marshall 1998). This announcement raised a number of fears, including the vision of research labs creating viable human/non-human chimeras.[†] The idea

---

*By the conventions of gene nomenclature, human genes are designated with all capital letters; mouse genes are written with only the initial letter capitalized. Thus *CDX2* and *Cdx2* are homologous genes found in humans and mice, respectively.

[†]A *chimera* is an organism that contains cells with different genetic constitutions (as compared to a normal organism, in which every cell in the body has the same genetic constitution). The term is drawn from Greek mythology, where it referred to a beast with a lion's head, a goat's body, and a serpent's tail.

offends the moral sensitivities of many, but there are those who support it as a method of obtaining human embryonic stem cells that would avoid many legal and ethical constraints. "Is there anything morally wrong with such research, if these cells are derived from somatic cells and never develop into embryos," asks Julian Savulescu, Director of the bioethics program at the University of Melbourne. "We now produce human proteins from human DNA inserted into animals. There does not seem anything objectionable about that. But if we can produce proteins in this way, why not blood cells?" (Savulescu 2000).

One of the dangers of using any non-human source of material, however, is the possible transfer of viruses across species. At present, most government agencies do not support any work that combines human nuclei with non-human oocytes (National Bioethics Advisory Commission 1999; Commission of the European Communities 2003). And since any such cells would have bovine mitochondria (because mitochondria are present in the egg cytoplasm, not in the nucleus), it is doubtful that any government agency would allow embryos generated with such eggs to be transplanted into humans.

Oocyte Donations

One of the moral issues raised by human embryonic stem cell research is that so many enucleated eggs are needed for such studies, and the collection of these poses unfair risks to women. During the egg collection procedure, a woman is given hormones to induce superovulation so that a large number of oocytes can be collected at one time. Among other risks (including cancer) associated with such therapy is the risk of a potentially life-threatening condition called ovarian hyperstimulation syndrome (see page 74). A mild form of this syndrome appears as a side effect in about 5 percent of all women who undergo ovarian hyperstimulation; 1–2 percent develop a severe, life-threatening form of the syndrome that can lead to kidney and lung failure, shock, or rupture of the ovary (Magee 2003).

In addition to the physical dangers, many people are concerned that poor women will be pressured into selling their oocytes to commercial suppliers. There are already cases (well documented in India; see, e.g., Jha 2004) where the very poor have sold organs such as their kidneys, undergoing major surgery and potential life-threatening complications to provide organs for wealthy recipients, in return for which the donors receive what are usually insultingly small amounts of money. Opponents feel that the projected need for vast numbers of eggs in stem cell research would open the door to exploitation of poor women.

Two recent discoveries may reduce the number of donor oocytes needed to be obtained from women. First, the number of oocytes needed to initiate stem cell lines may be significantly less than the numbers needed to establish viable cloned animals. Using freshly harvested oocytes from young women (rather than oocytes left over from fertility treatments), the South Korean laboratory of Hwang Woo-suk found that they could obtain stem cell lines from clones with transplanted nuclei with a success rate of 1 in 20 (Hwang et al. 2005). This is tenfold more efficient than previous attempts. In 9 of 11 attempts using fresh oocytes, stem cell lines from patient's nuclei succeeded with only a single donation.

Second, embryonic stem cells were shown to routinely form oocytes in culture and organize the cells around them into follicle cells, enabling the oocytes to mature normally (Hübner et al. 2003). Oocytes derived from embryonic stem cells rather than women's ovaries might therefore become a source of human oocytes for stem cell research. Ironically, this study raises other ethical issues about stem cells (Gilbert 2003): if they can produce normal oocytes, are they then totipotent rather than merely pluripotent, and if so, does a stem cell represent incipient life in the same way a zygote does? Could they or should they ever be used to create a viable embryo?

Life Extension and Age Retardation

The overriding objection of many people to embryonic stem cell research is their sense that it destroys a very early human life. One of the most profound issues raised by these potential therapies, however, has to do with the other extreme of life—senescence and death. What if stem cell therapy actually worked? What if it were so cheap and effective that if your heart cells needed repair, a simple trip to your physician's office or local hospital would be all you needed?

There have generally been two biological assumptions that could be made about our society. One is that the number of males and females born is relatively equal.* The second is that nobody lives productive lives past the century mark. We are now in the position to change the first assumption (see Chapters 5 and 6), and we may soon be in a position to change the second.

"Life extension" refers to increasing the number of years a person remains alive, and "age retardation" means slowing down the processes of senescence (the loss of mental and physical function as one gets older). Stem cells have the potential to do both. The staff of the President's Council on Bioethics has put forth a discussion paper that identified several possible outcomes of extending the healthy, productive lifespan. First, if knowledge of death gives our lives urgency and meaning, would a life removed from such considerations be less committed? Would we still want children? Indeed, since society is renewed by teaching its principles to children, would a society where so many of its members are aged be sustainable in the lack of such transmission?

---

*When the sex ratio is severely altered—as in time of war—dramatic social upheavals have occurred (see Jones 1980).

Moreover, society would have "glut of the able." Upper management need not retire, and professors need not quit their laboratories to make way for newcomers. One generation would have no need to make way for the next. This could lead to lack of innovation and adaptation (President's Council on Bioethics 2003). In an article entitled "The Coming Death Shortage," Charles Mann goes further, predicting that increased longevity would create an enormous economic crisis in which wealth would be concentrated almost exclusively in the hands of the healthy elderly. He maintains that "short of confiscating rich people's assets, it would be hard to avoid this divide between the elderly and everyone else" (Mann 2005).

As far back as 1969, Han Jonas noted that life-extending therapies might need to be restricted to the young—our society might have to say "no" to those above a certain age (Jonas 1969). Bioethicist David Callahan may speak for a growing number of people who believe that there is no societal need for increased longevity, and that medical progress would be better judged by its ability to help people achieve a peaceful and dignified death rather than averting death (Callahan 2003).

Public Policies Regarding Human Stem Cell Research

Interestingly, the intense controversy over human embryonic stem cell research has sometimes united disparate groups. Such "strange bedfellows" include politicians representing both pro-life and pro-choice positions (e.g., Senators Gordon Smith and Arlen Specter, both Republicans with pro-life stands unite with pro-choice senators in their support of using supernumerary IVF embryos for human stem cell research; Stolberg 2001), and religious groups normally with differing viewpoints (e.g., conservative Roman Catholics are aligned with Buddhist ethicists in their disapproval of stem cell research due to its destruction of potential life).

> [E]xtracting the stem cell from the embryo destroys the embryo, and this destroys its potential for life. Like a snowflake, each of these embryos is unique, with the unique genetic potential of an individual human being.
>
> U.S. PRESIDENT GEORGE W. BUSH (2001)

Many of the diseases that stem cell research holds the promise to alleviate are conditions that hit close to home for virtually all Americans. By adulthood, it is unusual to find anyone who hasn't been touched by friends or loved ones suffering from Parkinson's, Alzheimer's, or multiple sclerosis. The degenerative nature of these progressive diseases and their widespread and ubiquitous occurrence combine to make them especially frightening to many, and thus many people are torn: although they dislike the idea of destroying an embryo, they are eager to achieve the hoped-for relief the research tantalizingly holds out.

Although the federal government of the United States offers little support, stem cell research has moved forward with the support of numerous private research companies. Some state governments are in favor of promoting the research. In California, legislation passed early in 2005 earmarked $3 billion in state monies to fund stem cell research; the states of Massachusetts and New Jersey are also considering such legislative initiatives. Public visibility of the work has been boosted by the stories and activism of several celebrities who have been stricken. At least three major research efforts have been launched in association with high-profile Americans: the Christopher Reeve Paralysis Foundation, which funds research for developing cures for spinal cord injuries; the Michael J. Fox Foundation, which supports research for developing therapies for Parkinson's disease; and the Ronald and Nancy Reagan Research Institute, which promotes research to advance cures for Alzheimer's disease.

> Science has presented us with a hope called stem cell research, which may provide our scientists with many answers that for so long have been beyond our grasp. I just don't see how we can turn our backs on this. We have lost so much time already. I just really can't bear to lose any more.
>
> NANCY REAGAN (2004)

As both the debates and the research progress, it is society's responsibility to address these issues with policies that are in keeping with the moral conscience of the majority. In pluralistic societies such as the United States, this is difficult, especially when the problems being addressed involve highly technical details.

Government regulation

Several countries have determined that research on human embryonic stem cells should be regulated but vary widely on the level of regulation that they have instituted. Among countries that allow research with restrictions are Finland, Greece, the Netherlands, Sweden, Israel, and the United Kingdom. The United Kingdom, while having instituted strictly controlled regulations, is also among the most liberal in its government support of human embryonic stem cell research. For example, it was the first country to establish a government-supported human stem cell bank, accepting its first human cell lines derived from embryonic stem cells in May of 2004 (Pilcher 2004). More recently, government regulators in the United Kingdom granted the first one-year license to a laboratory, allowing it to pursue therapeutic cloning for work on a cure for Alzheimer disease (Timmons 2004). Other governments that support human embryonic stem cell research and therapeutic cloning (but not reproductive cloning) include Singapore, South Korea, and China. In China, the government has started building a state-run stem cell bank, including a transplant center and stem cell engineering development center.

Countries that presently prohibit the procurement of stem cells from human embryos include Germany, Austria, Denmark, France, Ireland, and Spain; these countries have varying policies on the use of human embryonic stem cell lines procured from other countries. Some of these are in the process of discussing revisions to their policies. A number of countries, including Belgium, Italy, Luxembourg, and Portugal, have no specific legislation. However, several of these nations are in the process of discussing regulation (Commission of the European Communities 2003).

In the United States and Canada, no federal laws regulate research on human embryos, though several laws are presently under discussion. In the United States, regulations are in the form of restrictions on use of federal funds for research. Citing his belief that blastocysts "have at least the potential for life," President George W. Bush has promised to veto any legislation that provides federal support for generating new stem cell lines. At present, research projects on human embryonic stem cells can only procure federal funding if they use human stem cell lines that existed as of August 9, 2001,

National Academies Guidelines for Stem Cell Research

Recognizing that stem cell research is going forward without government regulation, and that the unprecedented nature of the research opens new ethical questions, the scientific community has moved to address some of the issues by its own accord.

In April of 2005, the National Academies released detailed guidelines for researchers and institutions involved with stem cell research (http://national-academies.org). This group of private, nonprofit institutions provides science, technology, and health policy advice under a charter from the U.S. Congress.

Among the Academies' recommendations for conducting stem cell research are:

- Institutions involved in this research should establish Embryonic Stem Cell Research Oversight (ESCRO) committees to monitor experiments. These committees should include scientific, legal, and ethics experts as well as representatives of the general public.
- ESCRO committees are responsible for ensuring that full and informed consent is obtained from blastocyst donors. No payment for such donations is to be allowed. The committee is then

responsible for registering pertinent medical information about the donors and coding such information to ensure anonymity.

- Embryos should not be cultured for longer than 14 days (the point at which the body axes and neural tube begin to form).
- No nonhuman embryonic cells are to be transplanted into a human blastocyst. The transfer of human stem cells into nonhuman animals requires strong scientific justification and should be strictly monitored. Animals that have received infusions of human stem cells should not be allowed to reproduce.
- Researchers are not to pursue human reproductive cloning.

Richard Hynes, a cancer researcher at the Massachusetts Institute of Technology, co-chaired the group that wrote the guidelines. He sums up the feelings of many researchers in stating his view that "A standard set of requirements for deriving, storing, distributing, and using embryonic stem cell lines—one to which the entire U.S. scientific community adheres—is the best way for this research to move forward" (National Academies news release of April 26, 2005).

having been created from supernumerary IVF embryos. Though this source was supposed to include 78 stem cell lines, only 15 of these have been deemed usable (American Association for the Advancement of Science 2004). Moreover, cells from these 15 lines have been described by some researchers as "difficult to obtain, difficult to maintain, and poorly characterized" (Phimister and Drazen 2004).

Public or private funding?

Thus far, the United States government has forbidden the use of government funds for the creation of new stem cell lines and will not fund any stem cell research. (See Maienschein 2003 for a history of how this situation arose.) However, it has not outlawed stem cell research or the creation of new lineages; both of these activities continue, backed by private funding.

Thus, when Douglas Melton—a developmental biologist and the father of children with juvenile diabetes—wanted to study ways in which new insulin-secreting cells could be generated, he obtained funds not from government agencies but from private sources: the Juvenile Diabetes Research Fund, the Howard Hughes Medical Institute, and Harvard University. The stem cells Melton cultures were derived from frozen, unimplanted cleavage- and blastocyst-stage embryos obtained from a fertility clinic; the couples responsible for the embryos gave informed consent for this specific use of their cells. These cells have demonstrated the ability to differentiate into three germ layers and have a normal set of chromosomes in each cell (Cowan et al. 2004). Melton allows the stem cells lines generated in his laboratory to be made available to other researchers, without any personal financial gain.

The restrictions on federal funding for human embryonic stem cell research are a fiery issue. Those who feel the restrictions should be enhanced are generally opposed to any use of human embryos in research. Those who feel the restrictions should be relaxed argue that research in this country is advancing without the benefit of it being in the public realm. Because private companies usually keep the details of their experimental methods confidential, information that would be valuable to other scientists is often not passed on through peer-reviewed scientific publications; federally funded research, on the other hand, must be published, and information about the research is within the public realm from the time the grant application is approved.

In addition, private companies are under no obligation to follow the strict guidelines set up for federally funded research (although many companies voluntarily comply, setting up their own ethics committees to establish policy). Federally funded research is always under strict review and must abide by the rigid guidelines established by the granting agency, or the funding is revoked and disciplinary action may be taken (Dhanda 2002).

The opinions of the general public on this issue are split, but the positions of most scientific and medical associations (including the American

Society for Cell Biology, the Society for Developmental Biology, the American Medical Association, and the Coalition for the Advancement of Medical Research) are united in their support of relaxing the restrictions on federal funding. Most researchers would agree with Lawrence Goldstein (2000) of the American Society for Cell Biology, who testified before the U.S. Labor, Health and Human Services and Education Subcommittee that "stem cell research, in particular, has enormous potential for the effective treatment of human disease. Thus we believe that there is a moral imperative to pursue it in an ethically validated manner."

Patents

Registering a patent gives the discoverer or inventor of a product or process the exclusive right to their invention, so that no one can manufacture, make use of, or sell the product without the patent holder's permission. To date, over 200 patents have been granted for techniques and/or products involving embryonic stem cells, and at least 10 of those involve human stem cells (Scheinfeld and Bagley 2001). Does this use of patents, although established to encourage scientific progress by protecting the economic rights of people who develop new methods and products, have the effect of restricting access to the health care benefits that might develop from them?

Should human cells even be considered patentable subject material? The United States and the European patent offices have taken differing points of view on this issue. The U.S. Patent Office has issued patents for embryonic stem cell lines, while the European Patent Office has been unwilling to grant patents on stem cells (Whittaker 2003).

The three guidelines used by the U.S. Patent Office in granting patents are: novelty, inventiveness, and workability (the ability of the idea to actually be accomplished). The office does not specifically use morality as a guideline in refusing or allowing a patent; however, a patent request covering techniques to create a human-animal chimera was denied. The applicants, including researcher Stuart Newman and anti-biotechnology activist Jeremy Rifkin, wanted the patent in order to *prevent* the techniques from being used (Newman 2002). (Human-animal chimeras represent another new area of bioethical problems, especially when animals are given human brain tissue; see Shreeve 2005.)

The European office's refusal is based on the European Biotechnology Directive of 1998 that requires the exclusion from patenting of inventions that are "contrary to *ordre public* or morality" (Official Journal of the European Communities 1998). Those in disagreement with the European Patent Office stance state that this Directive confuses ethical concerns about science with those concerning patents (BioScience 2004). Conversely, there are Americans who would like to see moral concerns become part of the U.S. patenting criteria.

Justice: Distributing the Pie

If stem cell research eventually leads to cures, will these cures be equally available to all who could benefit from them? It is a fair assumption that such cures would be both technology-intensive and extremely expensive. Thus they would have limited availability in those countries without advanced medical systems; and in countries where health insurance is privatized, not all citizens have access to health care and expensive treatments usually are not distributed proportionately. Should a country such as the United States allow public money to be used to develop health treatments that will presumably benefit only the well-off, when this money could be put toward, say, expanded basic health care for the disadvantaged?

The realization that the treatments developed from human stem cell research may well benefit the wealthy over the poor leads some ethicists to argue against supporting this research. Because of the risks of diverting public funds away from the real health concerns of the poor, certain ethicists denounce the strong emphasis placed on stem cell research, agreeing with Suzanne Holland, professor of religious and social ethics at the University of Puget Sound, who writes, "[W]e would do well to subject public policy in question to the scrutiny of a moral litmus test that ensures the least well-off among us that they will be as likely to benefit as the most advantaged. Justice demands no less" (Holland 2001).

Counter to this argument is the sense among some medical researchers that the potential for this research is so great as to transcend any economic argument. Even if it were to initially benefit primarily the wealthy, the benefits that would accrue from alleviating these degenerative conditions—which affect vast numbers of people of all backgrounds—would eventually offset any early inequity.

The economic cost of maintaining victims of, for example, Parkinson's disease, is already extremely high. Caring for relatives with Alzheimer's disease likewise can devastate any family, but is far harder on those with fewer economic resources. Such conditions perhaps hit the many members of the "working middle class" even harder than the truly disadvantaged. The burdens of the caregivers in these situations—the majority of whom are women—should also be considered in the economic equation. In this vein, it seems to many that any therapeutic prospect with serious potential to lessen the impact of these long-term and widespread health problems should be vigorously pursued.

Human Dignity and the Moral Status of the Embryo

The moral debate about stem cells is not about good versus evil or science versus religion; it is about two competing notions of what is good for human dignity (see Gilbert 2001). The first concept of human dignity is an abstract notion maintaining that there is something special about being human that sets us apart from other animals. This "something special" could be our rationality (which encompasses our ability to articulate a concept such as "human dignity"), or our ability to communicate. The religious may view the "something special" as the human soul, that we are formed in God's image, or that only humans are extended the possibility of redemption.

One need not be religious to have the intuition that there is something special about being human; laws against slavery and cannibalism recognize the inherent worth of a human being above other animals. However, the religious notion of human dignity has on occasion been used to thwart improvements in the human condition. Conservative Christian groups vehemently opposed vaccination against smallpox, even a hundred years after its first use. Smallpox vaccine came from cows, and these groups felt that the injection of material from a cow into a human was an affront to human dignity.

The second concept of human dignity is more concrete. In this definition, part of our human dignity is found in the using of one's brain to ameliorate the consequences of disease (see Heschel 1985). Physicians often note that disease not only affects the body, but it can rob a person of his or her dignity. Thus, supporters of human stem cell research argue that such research has the potential to restore human dignity to the suffering. The Alzheimer's patient would be able to dress himself and recognize his family; the Parkinson's patient would be able to control her speech and movements; and the paraplegic would be able to walk and control his bowels.

> To save human life is to do the work of God. To heal is to do the holy … The act of healing is the highest form of *imatatio Dei* [acting as God would].
>
> RABBI A. J. HESCHEL (1985)

Supporters of stem cell research feel that it is more important to restore dignity to adult humans than to accord an abstract concept of human dignity to an embryo that has not yet become an individual (i.e., it can still form twins) and has no head, heart, arms, or even a distinguishable front or back. Thus, the Nobel Prize-winning biochemist Paul Berg told the United States Congress, stem cell research and clinical trials are important and "we are ethically and morally obligated to pursue them for the benefit of those who suffer" (Berg 2003). Similarly, embryologist John Gearhart feels that "it is the throwing away of fetal material that's unethical," since "its germ cells might be translated into lifesaving therapies" (Gearhart 2004). Thus, some religious groups, such as the Catholic Church, favor the first model of human dignity. Other religious groups, such as the Presbyterian Church and many Jewish groups, are in favor the second model.

The danger of this second vision of human dignity is that one can slide down a "slippery slope" to the point where any technological procedure that can be done should be done. As Leon Kass, chairman of the President's Council on Bioethics, wrote in 2001, "the real challenge of society is to find a way to reap the benefits of new biology without sliding down the road to *Brave New World* and human degradation."

Religious perspectives

When representatives of various religions discuss the ethical basis for allowing or disallowing the therapeutic use of human embryonic stem cells, most of the concerns center around the question, what is the moral status of the human embryo? Where disagreement arises is not in the general premise that human life is to be valued and protected, but rather is in the definition of human life and, more specifically, when life begins (Dhanda 2002). The views of various major religions on this point are discussed in Chapter 2. However, there are some points that are specific to the stem cell situation.

THE JEWISH PERSPECTIVE   From the Jewish perspective, the fertilized egg does not have "personhood" and the unimplanted embryo is without legal status. Jewish rabbis have frequently interpreted Genesis 9:6 ("Whoso shed-deth man's blood, by man shall his blood be shed") as being against abortion. However, since this text deals specifically with blood and the preimplantation embryo in a Petri dish has no blood, most feel these embryos can be used for study (Weiner 2005). Even if implanted, during the first 40 days these embryos are considered to be "as if they were water" (Dorff 1999). From this perspective, frozen embryos may be discarded or used for reasonable purposes, including stem cell procurement and subsequent stem cell research (National Bioethics Advisory Commission 1999). Moreover, the Jewish precept that humans have an obligation to heal means pursuing stem cell research in an effort to advance therapies is of great importance. Supernumerary embryos from IVF clinics are considered ethically usable for stem cell research. "In fact, such research might well be mandated to save life in Jewish law" (Zoloth 2001).

ROMAN CATHOLIC AND EASTERN ORTHODOX PERSPECTIVES   The Roman Catholic perspective that an embryo is an individualized human entity from the moment of fertilization effectively prohibits all human embryonic stem cell research. In this tradition, where abortion is totally unacceptable, using embryonic germ (EG) stem cells is also not permissible, since even using the discarded tissue from an aborted fetus to create these stem cells would constitute supporting abortion by the concept of "complicity" (National Bioethics Advisory Commission 1999).

In something of a contrast to the Roman Catholic tradition, the Eastern Orthodox Church views a human as progressing toward the likeness of God, and believes that this progression begins at fertilization. An elective abortion is seen as an act of defiance against God's grace. In this tradition, therefore, human embryonic stem cell research that uses supernumerary IVF embryos cannot be supported, since it implies complicity with abortion. However, because medicine is thought to be a divine gift and it is believed that humans have an obligation to heal, this tradition can support human embryonic stem cell research provided it uses already existing cell lines or stem cells obtained from miscarriages (National Bioethics Advisory Commission 1999).

CHRISTIAN PROTESTANT PERSPECTIVES   Christian Protestant perspectives are almost as diverse as the number of Christian Protestants. The most restrictive positions come from those who view the embryo as the weakest among us, requiring protection. This usually translates into a human embryo having equal moral status to that of an adult. From this perspective, any use of human embryos for research is unacceptable. Less restrictive positions support research on human embryonic stem cells obtained from supernumerary embryos from IVF clinics and fetuses aborted for therapeutic reasons (National Bioethics Advisory Commission 1999), and many Protestants eagerly support the research based on the perceived benefits it might engender.

ISLAMIC PERSPECTIVES   Islamic interpretations provide great tolerance for using human embryonic stem cells. It is thought that full human rights for an embryo do not occur until ensoulment, which most Islamic scholars consider occurs around 120 days after conception, though some view it as occurring as early as 40 days after conception. In this tradition, ethicists consider it acceptable to conduct research on human embryonic stem cells obtained prior to ensoulment, either from supernumerary IVF embryos or from embryos created specifically for research (National Bioethics Advisory Commission 1999).

The Common Disconnect between Ideal and Action

Most people will at times act in what can be seen as opposition to even deeply held moral convictions. Thus the senator who believes in the pacifist Christ will vote in favor of a war in which innocent children will undoubtedly be killed. The animal lover will be upset by the sight of dead deer in a hunter's truck but will go on to enjoy her hamburger for lunch.

It is perhaps necessary to human survival that most people are able to perform the mental gymnastics necessary to bridge the gap between their ideals and their actions. Examples of this "disconnect" can be drawn from any religious group. Damien Keown, editor of the *Journal of Buddhist Ethics* cites one from the primarily Buddhist country of South Korea. The opinions of Buddhist ethicists place great importance on not harming any life, human or non-human. Thus, whether or not the embryo is "human" is not of primary importance—it is a life. Moreover, the human life created at conception is believed to be the bearer of "the karmic identity of a recently deceased individual, [and] is therefore as entitled to the same moral respect as an adult human being."

Given this context, Keown points out that "it is interesting that Buddhists are the religious majority in the country where the latest breakthrough in stem cell research occurred." And despite the traditional Buddhist opposition to abortion, and despite the fact that abortion for social reasons is illegal as well as contrary to the majority religion, South Korea has been called an "abortion paradise"; figures of more than 1.5 million abortions performed yearly are often cited, suggesting that "there is unresolved dissonance between Buddhist teachings and practice on the moral status of embryonic life" (Keown 2004).

Stem cell research is a subject area where increased knowledge and experience may come to override political or religious convictions, especially because immediate and personally threatening health problems are involved. Many public figures in the United States who are against abortion rights and cloning in general have come out in favor of stem cell research for degenerative diseases. Nancy Reagan is probably the most prominent. Well known conservative legislators Orrin Hatch and the late Strom Thurmond also supported this research, despite their vehement vocal opposition to the right of a woman to end her pregnancy.

> Stem cell research facilitates life. Abortion destroys life. … The most pro-life position would be to help people who suffer from these maladies. That is far more ethical than just abandoning or discarding these embryonic stem cells.
>
> U.S. SENATOR ORRIN HATCH (2001)

Indeed, the readiness with which some conservative politicians have simultaneously espoused anti-abortion and pro-stem cell views has caused several philosophers and attorneys to speculate that the political "culture of life" in the United States is in fact primarily an attempt to erase the procreative and other legal rights women have gained since the 1960s and to return America to a place where men control both politics and the family. Such events as the recent assaults by anti-abortion pharmacists against a woman's right to obtain contraceptives (see page 58) and the derailment of U.S. Senate legislation that would fund family planning and teenage pregnancy prevention programs lend credence to this viewpoint (see Dogin 2004; Feldt 2004).

For those whose convictions are in fact rooted in faith rather than politics, the most difficult questions may arise if stem cell research—which is still in its infancy—eventually does produce viable cures for some of humanity's most feared conditions. As one bioethicist noted (Fiester et al. 2004), if it should become possible to cure your dog's diabetes with stem cells, how could you deny such benefits to your child or your father?

Would a person elect to watch their spouse suffer the deterioration of Alzheimer's disease because the cure for it was derived from embryonic cells? Would a paraplegic give up the opportunity to walk again on ethical grounds? If proposed stem cell therapies do not come to fruition, or if they prove ineffective or dangerous, the need to make such decisions will not arise, nor will society be presented with the huge ethical dilemma of how to decide who receives such therapies. Should stem cell research realize its hoped-for potential, however, individual suffering may be alleviated—but the effects this could have on the structure of society may present a new set of problems.

UNIT 6

Should We Modify the Human Genome?

Deliberate manipulation of the human germline will constitute a watershed in history, perhaps even in evolution. It should not be crossed surreptitiously, or before a full debate has allowed the public to reach an informed understand of where scientists are leading.

NEW YORK TIMES (JULY 22, 1982)

CHAPTER 11

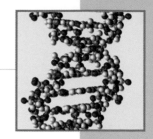

Gene Therapy

I n the twelfth century, the rabbi and physician Maimonides wrote that pious men of his day believed that an angel entered a woman's womb during each pregnancy and molded the material into an embryo. This, the men say, is a miracle. But, said Maimonides, how much greater a miracle it would be if God had so made matter that it did not need angelic assistance to form an embryo, but did so of its own accord.

Developmental biology asks the secular form of that question: How can matter form an embryo without external guidance? During each generation, the microscopic sperm and egg bring to the zygote all the genetic material needed to make an embryo. The egg also contains the cytoplasm, with its energy-supplying mitochondria. These genes provide the species-specific instructions to make heads, limbs, and organs out of the food provided by the mother.

This chapter will describe the fundamental principles of gene activation, and how our knowledge of genes and their activation has led to the possibility of biomedical modification of our genes, a field known as gene therapy. First we will examine the nature of genes and the genetic code.

The Anatomy of Genes: Codons and Amino Acids

Genes carry the instructions for making specific proteins. Proteins are the main component of the cells that make up our bodies, and it is these proteins that carry out the work of metabolism that keeps us alive. Some common proteins of the human body are listed in Table 11.1

Proteins are composed of amino acids—usually hundreds, sometimes even thousands of amino acids make up a single protein. Although there are

TABLE 11.1   Some proteins of the human body

Protein	Function
STRUCTURAL PROTEINS	
Myosin, actin	Allow muscles to contract
Collagen	Forms connective tissue (tendons, cartilage, ligaments) and supports the skin
Keratin	Stiffens skin; major component of hair and fingernails
ENZYMES	
Pepsin	Digestive enzyme; breaks food down into energy-providing chemicals
Adenosine deaminase (ADA)	Necessary for lymphocyte function; rids cells of toxic by-products
RNA polymerases	Initiate RNA construction from a DNA template
HORMONES AND ANTIBODIES	
Insulin	Regulates sugar metabolism
Erythropoeitin	Makes red blood cells
Human growth hormone (HGH)	Coordinates limb and skeleton growth
Immunoglobulins	Antibodies that protect the body against bacteria and viruses
TRANSPORT PROTEINS	
Hemoglobins	Carry oxygen from the lungs to cells
Cytochromes	Carry electrons in the transfer of energy within cells
Surfactants	Regulate water exchange between tissues (especially lung tissues) and air to prevent drying out.

TABLE 11.2   The 20 amino acids and their abbreviations

Amino acid	Abbreviations 3-letter	Abbreviations 1-letter	Amino acid	Abbreviations 3-letter	Abbreviations 1-letter
Alanine	Ala	A	Leucine	Leu	L
Arginine	Arg	R	Lysine	Lys	K
Asparagine	Asn	N	Methionine	Met	M
Aspartic acid	Asp	D	Phenylalanine	Phe	F
Cysteine	Cys	C	Proline	Pro	P
Glutamic acid	Glu	E	Serine	Ser	S
Glutamine	Gln	Q	Threonine	Thr	T
Glycine	Gly	G	Tryptophan	Trp	W
Histidine	His	H	Tyrosine	Tyr	Y
Isoleucine	Ile	I	Valine	Val	V

many individual amino acid units in a protein, there are in fact only 20 amino acids (Table 11.2); arranging these 20 amino acids in an incredibly large number of different sequence combinations gives each protein different and highly specific properties.

The amino acid sequence that specifies a given protein is encoded in the gene for that protein. Genes are chains of **DNA (deoxyribonucleic acid)** molecules. The DNA is made up of paired molecules called nucleotides that hold the molecular code for the 20 amino acids. In order for the code to "work," the nucleotide pairs separate into two strands, and one of them—the **template**—creates a single-stranded copy that is an **RNA (ribonucleic acid)** molecule. It is this messenger molecule, or **mRNA**, that can create a protein.

Figure 11.1 is a simple summary of how the DNA molecules in genes become proteins. Let's take a closer look at the players in this "central dogma."

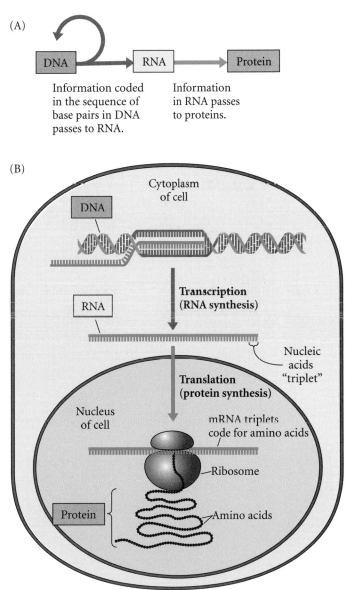

FIGURE 11.1 From gene to protein. (A) The "central dogma" of molecular biology. DNA can replicate (make more of itself) and sequences of DNA can be transcribed into RNA. The RNA can be translated into protein. (B) Expansion of the flow chart in (A). The properties of the protein are determined by its sequence of amino acids. The sequence of amino acids is determined by the sequence of nucleic acids in the messenger RNA (mRNA), and the sequence of nucleotides in the mRNA is determined by the sequence of nucleotides in the chromosomal DNA.

Transcribing the nucleotides: From DNA to RNA

Just as proteins are composed from 20 amino acids, the DNA and RNA that specify the amino acids are made up of four **nucleotides**. The central element of each nucleotide is a nitrogen-containing **base**. The four bases that are used in DNA are **adenine (A)**, **cytosine (C)**, **guanine (G)**, and **thymine (T)**. RNA is also made up of four bases; the first three (A, C, and G) are the

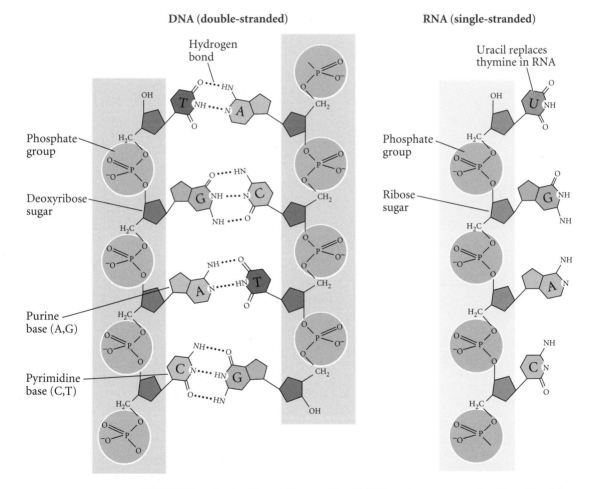

FIGURE 11.2   Nucleotides and base pairing. (*Left*) Complementary nucleotide pairing in DNA. Each adenine (A) is paired with a thymine (T) and each guanine (G) is paired with a cytosine (C) to form the "rungs" on a ladder. The deoxyribose sugars of these nucleotides are connected through phosphate groups to form the "ladder's" two sides. The cell's genes are encoded in the DNA. (*Right*) RNA is a single strand of nucleotides. It is created as the complement of one of the two DNA strands; uracil (U) replaces thymine in RNA. RNA is the messenger that "reads" the coded DNA strands so that they can be translated into proteins.

same, but in RNA thymine is replaced with a slightly different molecule, **uracil (U)**. Each nucleotide also contains a sugar molecule (deoxyribose sugar in DNA; ribose sugar in RNA) and a phosphate group (Figure 11.2).

As can be seen in Figure 11.2, adenine and guanine (the **purines**) are slightly larger than cytosine and thymine or uracil (the **pyrimidines**). This size difference is crucial to the principle of **complementary base pairing** in DNA: in order to keep the "rungs" of the DNA "ladder" at equal widths, a purine must always pair with a pyrimidine. The base pairing rules are in fact even more specific: in DNA, an A always binds to a T (or a T to an A), and a G always binds to a C (or vice versa). The nucleotide "ladder" formed by the base pairs twists to form the famous DNA "double helix" that has become the icon of modern biology (Figure 11.3).

DNA stays inside the cell's nucleus. During cell division, the two strands come apart and a new strand is synthesized onto the template of the existing one. This is how the chromosomes duplicate themselves during mitosis.

To transform its chain of nucleotide instructions into proteins, the DNA has to make a single-stranded copy of itself, and this copy is the **messenger RNA (mRNA)**. To initiate the production of mRNA, a set of enzyme proteins called the RNA polymerase complex opens up the double strand of DNA and binds to one of the strands (the template). It then proceeds along that strand, using nucleotides present in the nucleus to create a pre-mRNA

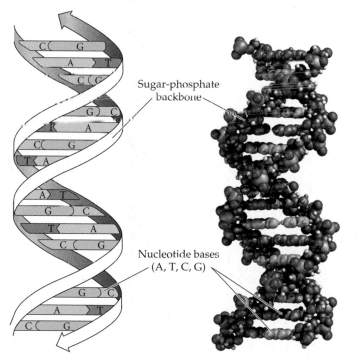

Sugar-phosphate backbone

Nucleotide bases (A, T, C, G)

FIGURE 11.3 The two strands of DNA go in opposite directions and twist to form a "double helix."

FIGURE 11.4 The road map from nucleus to cytoplasm in human cells. The transcribed RNA is not an mRNA, but a pre-mRNA that contains both exon and intron sequences (see page 187). The introns are processed out of the pre-mRNA and the mRNA is released into the cytoplasm to be translated into protein by the ribosomes.

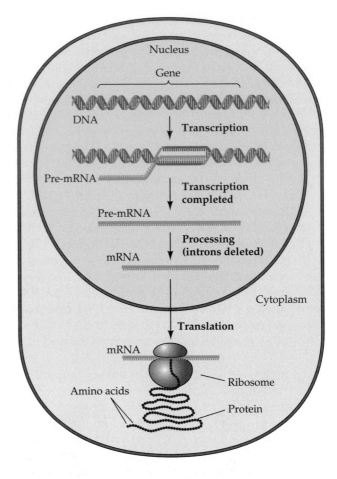

strand that is complementary to the template DNA strand. For every G in the DNA template, RNA polymerase inserts a C in the pre-mRNA strand. For every C in the DNA, it inserts a G. For every A it inserts a U, and for every T it inserts an A. This process, called **transcription**, results in a single long string of nucleotides that carry a specific message. The pre-mRNA is then processed by deleting certain sequences, and the resulting messenger RNA is then transported from the nucleus into the main part of the cell, the cytoplasm (Figure 11.4), where the message will be decoded into the amino acid sequence for a specific protein.

Translating the nucleotides: From RNA to protein

In the cytoplasm, the mRNA binds to the protein-synthesizing apparatus of the cell, the **ribosomes**. Ribosomes are complexes made up of over a hun-

Maternal Messenger RNAs: Getting Things Started

RNA polymerase and ribosomes are necessary in order to translate mRNA into proteins. But these enzymes are themselves proteins that need to be transcribed and translated. So where do the polymerase and ribosomes that translate the *first* RNA polymerase and ribosomes in the embryo come from? This is a true "chicken-or-egg" problem, especially since the solution turns out to involve the egg.

As it matures in the ovary, the oocyte, using genes present in its own genome, synthesizes enormous amounts of ribosomes and mitochondria. Moreover, the developing oocyte synthesizes and stores numerous different types of proteins and mRNAs. These stored proteins and mRNAs include

RNA polymerase; the tubulin and actin proteins that form the spindles needed for cell division; and the proteins needed for DNA replication.

Many of these stored proteins and mRNAs have molecular inhibitors attached to them so that they are not actually expressed in the oocyte. However, the events of fertilization (especially the release of calcium ions that happens when the sperm and egg fuse; see Figure 3.9) release the inhibiting molecules so the proteins can function and the mRNAs can be translated. Thus, the first proteins needed for transcription and translation are made by the mother's genome and are already packaged in the egg, ready to be released and used as soon as fertilization takes place.

dred proteins combined with three major types of RNA (called ribosomal RNAs, or rRNAs, to distinguish them from messenger RNA). Ribosomes "read" the nucleotides of the mRNA strand in groups of three. Three nucleotides encode one amino acid (the famous "genetic code" shown in Figure 11.5), and the ribosomes build proteins using this information. In other words, any ribosome can make any type of protein; which protein they make is determined by the mRNA that binds to them. This process by which ribosomes make a protein based on the sequence of nucleotides in mRNA is called **translation** (see Figures 11.1B and 11.4).

THE GENETIC CODE   Based on the four nucleotide bases of RNA (A, U, C, and G), there are 64 possible combinations of three bases each, called **nucleotide triplets** or, more often, **codons**. Most of the triplets encode amino acids, and since there are more codons (64) than there are amino acids (20), several codons can code for the same amino acid (Figure 11.5).

There are three codons that do not encode amino acids. Rather, they encode a "stop" signal. When the ribosome encounters such a codon, it is told that the protein is finished and should be released.

The elucidation of the genetic code took many years, coming to fruition over the course of the 1960s. The code is without doubt one of the most important discoveries in modern biology. It appears that all living organisms use the same genetic code; this fact is one of the strongest pieces of evi-

Second letter

	U	C	A	G	
U	UUU / UUC Phenyl-alanine UUA / UUG Leucine	UCU / UCC / UCA / UCG Serine	UAU / UAC Tyrosine UAA Stop codon / UAG Stop codon	UGU / UGC Cysteine UGA Stop codon / UGG Tryptophan	U / C / A / G
C	CUU / CUC / CUA / CUG Leucine	CCU / CCC / CCA / CCG Proline	CAU / CAC Histidine CAA / CAG Glutamine	CGU / CGC / CGA / CGG Arginine	U / C / A / G
A	AUU / AUC Isoleucine AUA AUG Methionine; start codon	ACU / ACC / ACA / ACG Threonine	AAU / AAC Asparagine AAA / AAG Lysine	AGU / AGC Serine AGA / AGG Arginine	U / C / A / G
G	GUU / GUC / GUA / GUG Valine	GCU / GCC / GCA / GCG Alanine	GAU / GAC Aspartic acid GAA / GAG Glutamic acid	GGU / GGC / GGA / GGG Glycine	U / C / A / G

First letter (left column) · Third letter (right column)

FIGURE 11.5   The genetic code. Common to all life on Earth, the triplet code of the mRNA specifies the amino acid sequence of the protein. Each three-letter unit of the mRNA (the codon) is "decoded" by the ribosome into an amino acid. To decode the codon, find its first letter in the left hand column, then read across the top for the second letter, then read down the right-hand column for the third letter. For example, the codon AUG in the mRNA corresponds to the amino acid methionine in the protein. (This codon is also the one that always starts the coding region of the mRNA).

dence for the evolutionary origin of all life from a common ancestor. Indeed, ribosomes from human cells can make proteins using the mRNA from bacteria (and vice versa).

MUTATIONS   When the sequence of nucleotide bases encoded in the DNA or mRNA is altered, we say that a **mutation** has occurred. Mutations can be mistakes—"typographical errors," if you will—in transcription or translation. They can also be the result of subtle biochemical reactions within the cell. Or they can be caused by outside agents—**mutagens**—such as radiation, toxins, or chemicals (nicotine, for example).

Many mutations have no effect on the protein coded for in the DNA. For example, as you can see from Figure 11.5, if the final C in the triplet ACC mutated and became a U, the resulting triplet—ACU—would still code for the amino acid threonine, so the protein would be unchanged. But if the A in ACC mutated into a G, the triplet GCC would code for alanine, which has a different biochemistry than threonine. Depending on where in the chain it occurred, such a mutation could have a major effect on the protein's function.

For instance, on human chromosome 4 there is a gene that encodes a receptor protein for a particular growth factor. If the G at position 1138 of the mRNA for this growth factor gene is mutated into an A, the codon changes from GGA to AGA. This in turn changes the amino acid at position 380 from glycine—a small molecule with no electric charge—to arginine, a much larger molecule that carries a negative electric charge. This substitution, at that particular position, drastically alters the structure and activity of the receptor protein (Rousseau et al. 1994; Shiang et al. 1994). The end result is that the arm and leg bones of a person with this mutation stop growing much too early, and the person is very short in stature. This condition is a type of dwarfism called achondroplasia.

Similarly, if a mutation converts an early-occurring triplet into one of the three termination codons (see Figure 11.5), the protein is immediately released from the ribosome, long before it is finished. Such a protein will almost certainly be nonfunctional, and the consequences (depending on the protein's function) can be lethal.

The Anatomy of Genes: Exons and Introns

The first genes to be sequenced were those of the bacterium *Escherichia coli*, a common laboratory organism obtained from the human gut. In *E. coli* genes and in those of other bacteria (single-celled organisms with no nucleus), the gene sequences are linear and unbroken: every codon reflects an amino acid in the protein sequence. As more complicated genes were sequenced over the course of the 1970s, scientists made a totally unexpected discovery. In eukaryotic organisms (those organisms whose cells have nuclei: protists, plants, fungi, and animals) the regions of genes that are actually translated into amino acids are interrupted by nucleotide sequences that do not produce amino acids. The amino-acid producing sequences of a gene were dubbed **exons**, while the intervening non-coding sequences were called **introns**.

For instance, one of the protein subunits of hemoglobin* contains 146 amino acids (Figure 11.6). The first 30 amino acids of this protein are encoded together on 120 nucleotides of DNA. Then there is an intervening sequence of 130 nucleotides that do not encode a protein. This intron is followed by another stretch of nucleotides that encode 74 amino acids, followed by a longer intron (850 nucleotides), and finally an exon that encodes amino acids 104 through 146. Immediately following the codon for the last amino acid is an UAA codon that terminates the protein.

---

*Hemoglobin is made up of four separate subunits—two molecules each of the alpha-globin and beta-globin (α-globin, β-globin) proteins. Each of the four protein subunits is bound to an iron-containing heme group that carries a molecule of oxygen.

Enhancer   Enhancer   Promoter   Translation

ccctgtggagccacaccctagggttggccaatctactcccaggagcagggagggcaggagccagggctgggcataaaa   initiation site

gtcagggcagagccatctattgcttACATTTGCTTCTGACACAACTGTGTTCACTAGCAACCTCAAACAGACACCATG   (methionine)

ValHisLeuThrProGluGluLysSerAlaValThrAlaLeuTrpGlyLysValAsnValAspGluValGlyGlyGlu   Exon 1
GTGCACCTGACTCCTGAGGAGAAGTCTGCCGTTACTGCCCTGTGGGGCAAGGTGAACGTGGATGAAGTTGGTGGTGAG

AlaLeuGlyArg
GCCCTGGGCAGGTTGGTATCAAGGTTACAAGACAGGTTTAAGGAGACCAATAGAAACTGGGCATGTGGAGACAGAGAAG

                                                      LeuLeuValValTyr
ACTCTTGGGTTTCTGATAGGCACTGACTCTCTCTGCCTATTGGTCTATTTTCCCACCCTTAGGCTGCTGGTGGTCTAC

ProTrpThrGlnArgPhePheGluPheGlyAspLeuSerThrProAspAlaValMetGlyAsnProLysValLys   Exon 2
CCTTGGACCCAGAGGTTCTTTGAGTCCTTTGGGGATCTGTCCACTCCTGATGCTGTTATGGGCAACCCTAAGGTGAAG

AlaHisGlyLysLysValLeuGlyAlaPheSerAspGlyLeuAlaHisLeuAspAsnLeuLysGlyThrPheAlaThr
GCTCATGGCAAGAAAGTGCTCGGTGCCTTTAGTGATGGCCTGGCTCACCTGGACAACCTCAAGGGCACCTTTGCCACA

LeuSerGluLeuHisCysAspLysLeuHisValAspProGluAsnPheArg
CTGAGTGAGCTGCACTGTGACAAGCTGCACGTGGATCCTGAGAACTTCAGGGTGAGTCTATGGGACCCTTGATGTTTT

CTTTCCCCTTCTTTTCTATGGTTAAGTTCATGTCATAGGAAGGGGAGAAGTAACAGGGTACAGTTTAGAATGGGAAC

AGACGAATAGATTGCATCAGTGTGGAAGTCTCAGGATCGTTTTAGTTTCTTTTATTTGCTGTTCATAACAATTGTTTTC

TTTTGTTTAATTCTTGCTTTCTTTTTTTTTTCTTCTCCGCAATTTTTACTATTATACTTAATGCCTAGTACATTACTATT

AACAAAAGGAAATATCTCTGAGATACATTAAGTAACTTAAAAAAAAAACTTTACACAGTCTGCCTAGTACATTACTATT

TGGAATATATGTGTGCTTATTTGCATATTCATAATCTCCCTACTTTATTTTCTTTTATTTTTAATTGATACATAATCA

TTATACATATTTATGGGTTAAAGTGTAATGTTTTAATATGTGTACACATATTGACCAAATCAGGGTAATTTTGCATT

TGTAATTTTAAAAAATGCTTTCTTCTTTTAATATACTTTTTTGTTTATCTTATTTCTAATACTTTCCCTAATCTCTTT

CTTTCAGGGCAATAATGATACAATGTATCATGCCTCTTTGCACCATTCTAAAGAATAACAGTGATAATTTCTGGGTTA

AGGCAATAGCAATATTTCTGCATATAAATATTTCTGCATATAAATTGTAACTGATGTAAGAGGTTTCATATTGCTAA

TAGCAGCTACAATCCAGCTACCATTCTGCTCTTTTATTTTATGGTTGGGATAAGGCTGGATTATTCTGAGTCCAAGCTAG

                                          LeuLeuGlyAsnValLeuValCysValLeuAla   Exon 3
GCCCTTTTGCTAATCATGTTCATACCTCTTATCTTCCTCCCACAGCTCCTGGGCAACGTGCTGGTCTGTGGCTAATGCCCTG

HisHisPheGlyLysGluPheThrProProValGlnAlaAlaTyrGlnLysValValAlaGlyValAlaAsnAlaLeu
CATCACTTTGGCAAAGAATTCACCCCACCAGTGCAGGCTGCCTATCAGAAAGTGGTGGCTGGTGTGGCTAATGCCCTG   Translation
                                            termination site
AlaHisLysTyrHis
GCCACACAAGTCACTAAGCTCGCTTTCTTGCTGTCCAATTTCTATTAAAGGTTCCTTTGTTCCCTAAGTCCAACTAC

TAAACTGGGGGATATTATGAAGGGCCTTGAGCATCTGGATTCTGCCTAATAAAAAACATTTATTTTCATTGC

FIGURE 11.6   The DNA sequence of the human gene for β-globin, an oxygen-carrying protein found in the hemoglobin of red blood cells. The three coding sequences (exons) are shown in colors, with the amino acids encoded by each "triplet" indicated directly above. The exons are separated by introns (gray). Introns are nucleotide sequences in the DNA that do not encode amino acids. Introns are are spliced out of mRNA as it is processed. Before the ATG codon (blue) that initiates translation, there is a "leader sequence" (lowercase letters) containing enhancers and a promoter (see page 190).

Therefore, in mammals (as in other eukaryotes), the first RNA that gets made is not a mature mRNA but rather a **pre-mRNA** that contains the intervening (intron) sequences. This pre-mRNA is processed within the nucleus by removing the intervening sequences. As it is processed, it pass-

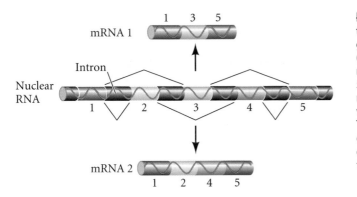

FIGURE 11.7   Differential processing of the pre-mRNA in different cell types can give different mRNAs. In one cell type (top), regions 2 and 4 of the pre-mRNA are recognized as introns, and regions 1, 3, and 5 are seen as exons. In cell type 2 (bottom), regions 1, 2, 4, and 5, are seen as exons, while region 3 is perceived to be an intron. Thus mRNAs 1 and 2 will be translated into different proteins. Differential processing explains how the same gene can give rise to different proteins in different cell types.

es from the nucleus through the nuclear pores and into the cytoplasm. The mRNA that binds to ribosomes is a mature mRNA (that is, an mRNA with no introns) that can be translated into a linear amino acid sequence with no introns (see Figure 11.4).

The presence of intervening sequences in genes has profound implications. It has recently been discovered that some cells recognize a particular nucleotide sequence as an intron (and thus remove it from the mRNA), while other cells recognize the *same sequence* as an exon and retain it in the mRNA, where it encodes a portion of a protein. Thus, the same gene can generate many different mRNAs, and each of these mRNAs makes a different protein (Figure 11.7). This **differential processing** of mRNA is probably the main reason why the approximately 20,000 human genes can encode as many as 5 to 10 times that many proteins. In other words, the human **genome** (total number of genes) is much smaller than the human **proteome** (the total number of proteins that the genome is capable of synthesizing).

The Anatomy of Genes: Promoters, Enhancers, and Insulators

In addition to the structural portion of the gene—the exons that are transcribed to form a protein—there are regions of the gene that regulate which exons are transcribed, and when, where, and for how long. These regulatory regions include the promoter, enhancers, and insulators.

- The **promoter** is a region of DNA containing a specific sequence that binds the RNA polymerase so it can unwind the double helix and initiate the synthesis of mRNA.

- **Enhancers** are sequences that bind transcription factors. The resulting complexes tell the gene when to be active, where to be active, and how much RNA product it should synthesize.

- **Insulator elements** prevent the enhancer's ability to stimulate gene activation from spilling over into neighboring genes.

The promoter sequence

Every gene has a **promoter**. The DNA sequence of the promoter tells the RNA polymerase to bind to it and start the gene's transcription. The promoter also orients the polymerase so that it "reads" in the correct direction. This makes sense, since if RNA polymerase could bind anywhere on the DNA, mRNAs would have no way of defining the starting point of the protein. The promoters can thus be thought of as a punctuation mark that tells RNA polymerase (the "reader") where to start. Once RNA polymerase binds to the promoter and transcription begins, the gene is activated.

Enhancers and differential gene activation

Every cell contains the same genes, but different mRNAs are transcribed in each different cell type. Indeed, one of the most important concepts in developmental biology is **differential gene activation**. That is to say, certain genes are active only in certain cell types. The hemoglobin genes are active only in those cells that will become red blood cells. The insulin gene is active only in certain cells of the pancreas. The genes encoding the enzymes that make the melanin pigment of our skin are active only in the melanocytes that migrate to our epidermis. How can a gene be active in one cell type and not in another?

The answer is the **enhancer**. Enhancers are regions near the gene that, when stimulated by binding transcription factors (described in Chapter 9; see Figure 9.3), tell the gene when it can be active, where it can be active, and how much RNA it can make. The globin genes, for instance, have enhancers that are only activated in red blood cells. Specific **transcription factors** bind to a particular enhancer region of DNA and activate the globin gene associated with that enhancer. In another example, the somatostatin gene is activated only in gut tissues by gut-specific transcription factors (Figure 11.8). This activation is usually accomplished by allowing the RNA polymerase complex to bind efficiently to the promoter. Enhancers can bind several transcription factors, and it is the specific combination of transcription factors present that allows a gene to be active in a particular cell type. That is, the same transcription factor, in conjunction with different other transcription factors, will activate different promoters in different cells.*

Insulators

Because the same enhancer may be able to interact with several different promoters, a gene is often flanked by DNA sequences called insulators.

---

*In Chapter 7, we described how methylation of DNA prevents a gene from becoming active (see pages 122–123). Here we can see why: the methylated DNA is usually in the enhancer region; the transcription factors cannot bind to methylated DNA, so the gene remains inactive.

FIGURE 11.8   The enhancer region for the gene encoding somatostatin. Somatostatin is a protein made in the pancreas and intestine, and it is involved in growth control. Pdx1, Pbx1, Pax6, and CREB are all transcription factors that bind to the enhancer and interact with the RNA polymerase complex to bind the polymerase to the promoter. It is this combination of transcription factors that causes the somatostatin gene to be activated.

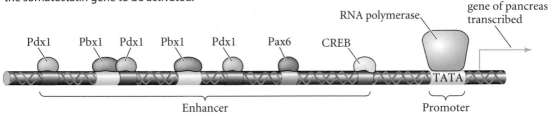

These insulator regions bind proteins that prevent the enhancer's ability to stimulate gene activation from spilling over into a neighboring gene. Insulators can also work in the opposite way, by protecting a gene from being *in*activated.

Insulators are currently the least understood of the gene's regulatory elements, but it is clear they are essential in defining the boundaries of a gene's expression (hence they are sometimes known as "boundary elements"). Experiments with mice have indicated that if certain insulator regions are inactivated, the embryo's gene expression pattern is faulty and the embryo dies (Hartwell et al. 2004).

Gene Therapy

In order for gene therapy to be successful, we have to know the locations of both the structural sequence of the gene and the regulatory sequences that affect their transcription. One also has to know which transcription factors are needed to activate the gene, and whether the cells being treated contain the appropriate transcription factors.

Somatic cell gene therapy

Somatic cell gene therapy raises the hopes of curing a person's genetic disease by the insertion of a wild-type (normally functioning) gene that would be activated at the appropriate times and places. In such cases, committed stem cells from the patient (such as blood stem cells or liver stem cells) are removed and cultured. The normal gene is inserted into the cells' nuclei and the cells are then reinserted into the patient. The wild-type gene is "packaged" in a vector (carrier)—usually a harmless virus—that enables the new gene to enter the cell and be transcribed (Figure 11.9). This type of gene therapy is currently being tested in over 600 laboratories.

FIGURE 11.9   Somatic cell gene therapy. A cloned wild-type (normally functioning) gene is "packaged" in a harmless virus (the vector, or carrier) and then inserted into the nuclei of cultured stem cells (such as blood stem cells) from a patient with a deficient gene; the "engineered" cells are then injected back into the patient, where the added gene will restore normal function.

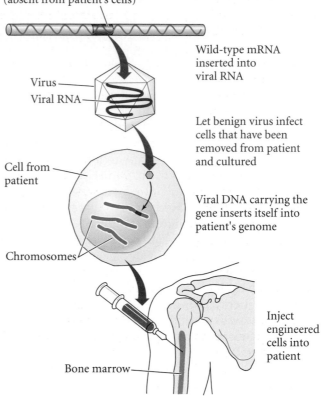

Cloned wild-type gene
(absent from patient's cells)

Wild-type mRNA
inserted into
viral RNA

Virus
Viral RNA

Let benign virus infect
cells that have been
removed from patient
and cultured

Cell from
patient

Viral DNA carrying the
gene inserts itself into
patient's genome

Chromosomes

Inject
engineered
cells into
patient

Bone marrow

Somatic cell gene therapy has worked well in some situations. In 1990, W. French Anderson and his colleagues treated several young children suffering from severe immunodeficiency. The children had a deficiency in the gene for the enzyme adenosine deaminase (ADA), which is normally expressed in T-cells, a type of white blood cell that is crucial to the body's immune response. Without functioning ADA, a person's ability to fight infection is reduced to the point where even a simple cold may be lethal.

In the treated children, a sample of their own (*ADA*-deficient) immature lymphocytes and lymphocyte stem cells was cultured. The functioning *ADA* gene was inserted into the genome of a harmless virus, which was then allowed to "infect" the cultured precursor cells. These cells incorporated the corrected gene and were injected back into the patient. The treatment has been successful in several cases; the patients' immune system function was restored, although they undergo continual medical treatment (Blaese et al. 1995).

Somatic cell gene therapy has also been used to successfully treat genetic liver disease and hemophilia (Roth et al. 2001). However, in 1999, an 18-year-old man undergoing gene therapy treatment for a genetic, kidney-specific enzyme defect died four days after beginning treatment. The cause of his death remains a mystery, but it is possible that the virus vector stimulated an allergic-type reaction that shut down the patient's lungs (Stolberg 1999; Couzin and Kaiser 2005). Similarly, a promising study concerning an inherited immune deficiency similar to the *ADA* deficiency was halted when three of the patients developed leukemia-like symptoms (Check 2005).

There are always safety concerns when a new medical technology is introduced to the market. No technology is ever completely safe, and different people respond differently to different chemicals. The safety issues maybe technical ones that will vanish as the techniques of gene therapy become more thoroughly tested. However, safety remains an important issue and we should be cautious about expecting "too much, too soon" from somatic cell gene therapy.

Indeed, unrealistic expectations generated by media "hype" may be one of the biggest problems faced by somatic cell therapy. It is somewhat reminiscent of the early 1900s, when radium was touted by the media as the cure for cancer. Radiation therapy indeed became an indispensable part of cancer treatment, but it is not in and of itself a cure. Will gene therapy mean that Parkinson's disease, Alzheimer's disease, diabetes, and cystic fibrosis will all become curable within the next few years? Probably not, since techniques are still being worked on, and this will take time.

Some diseases will be more readily cured by somatic cell gene therapy than others. For instance, even though we know (down to the nucleotide level) exactly which gene is responsible for the deadly condition sickle cell anemia, and we know that the gene must be expressed in red blood cells, and we know how to modify red blood cells to take up new genes, we still have no genetic therapy for sickle cell anemia (Holtzman and Marteau 2000). One of the reasons is that sickle cell anemia is a *gain*-**of-function mutation**. The genes for the hemoglobin proteins in sickle cell anemia have mutated to give the proteins a new ability—they stick to one another. These "sticky" proteins eventually form cables that give the mutant blood cell its sickled shape and limit its flexibility, eventually resulting in clogged blood vessels and early death. It is easier to cure a *loss*-**of-function mutation**, in which the functioning gene is simply missing. In loss-of-function, one can *add* a functioning gene where there was no functioning gene; in treating gain-of-function, you have to get rid of the mutant gene and then *substitute* the wild-type, functioning gene.

The genetic engineering procedures described above are essentially standard, albeit experimental, medical treatments wherein an individual is being treated for a life-threatening condition; thus the ethical decisions to be made are akin to those involving human tests on any new drug or sur-

Transgenic Embryonic Stem Cells

The combination of somatic cell nuclear transplantation and gene therapy can be used to cure *genetic* diseases that affect specific organs or tissues. For instance, mice and humans deficient in the *Rag2* gene are unable to recombine their DNA to make antibodies. Patients with mutations in *Rag2* have a genetic immunodeficiency syndrome wherein they cannot protect themselves against many infectious diseases.

Researchers took tail-tip cells from a *Rag2*-deficient mouse and implanted the nuclei into enucleated mouse oocytes (Rideout et al. 2002). The renucleated oocytes were activated and gave rise to

REPAIR OF A DEFECTIVE GENE by therapeutic cloning in mice. Tail tip cells from a *Rag2*-deficient mouse are cultured, and the nuclei of these somatic cells are transplanted into activated enucleated oocytes. The oocytes develop to the blastocyst stage, and the *Rag2*-deficient inner cell mass cells are isolated and grown as stem cells. These stem cells are then incubated with wild-type (i.e., normal) *Rag2* DNA and homologous recombination allows some of these cells to replace the mutant allele with a wild-type allele. The repaired ES cells are aggregated, and given factors that direct them to differentiate into hematopioetic and lymphocytic stem cells. These cells are injected into the mouse and can restore its immune system with *Rag2*-positive cells that can make antibodies.

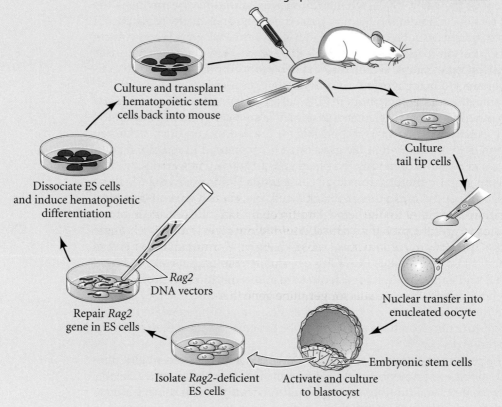

Culture and transplant hematopoietic stem cells back into mouse

Culture tail tip cells

Dissociate ES cells and induce hematopoietic differentiation

Rag2 DNA vectors

Repair *Rag2* gene in ES cells

Nuclear transfer into enucleated oocyte

Embryonic stem cells

Isolate *Rag2*-deficient ES cells

Activate and culture to blastocyst

Transgenic Embryonic Stem Cells (continued)

blastocysts, from which *Rag2*-deficient ES cells were cultivated. One of the mutant *Rag2* genes was then converted into a wild-type gene by homologous DNA recombination, and the "repaired" embryonic stem cells were grown and cultured to produce hematopoietic stem cells. These hematopoietic stem cells were injected back into the mouse from which the cells were originally taken. These genetically repaired cells repopulated the mouse's bone marrow, and mature antibody-producing cells were detected within a month of the transplant. Thus, it appears

some genetic disorders may be treatable by a combination of gene therapy and therapeutic cloning.

In another study, researchers inserted a mutant gene for myostatin (a muscle-forming gene) into mouse embryonic stem cells and injected these transgenic stem cells into normal mouse blastocysts. The mutant gene was known to stimulate increased muscle formation, and the mice that developed from these blastocysts did indeed have 5–20 percent more muscle mass than did their nontransgenic littermates (Pirottin et al. 2005).

---

gical procedure in market-driven economies. However, there is a very fine line between treatment and *enhancement*. The drug finasteride, for example, stops the prostate from enlarging and is medically prescribed to men with this serious condition. In smaller doses, this same drug will retard male pattern baldness. (Both prostate growth and pattern baldness result from the metabolizing of testosterone to a more potent steroid hormone, and finasteride blocks that metabolism.) Those men who can afford the drug can impede baldness by using it, and it is widely marketed for this purpose under the brand name Propecia[R].

Applying such standards to gene therapy, the same type of treatment that could cure genetic diseases and cancers might also be used to alter the phenotype of a "normal" individual. Inserting certain genes into muscles, for instance, could make a person stronger or less likely to experience muscle fatigue (see pages 200–201 and Pirottin et al. 2005). It is possible genes might be identified that, if modified, could allow a person to live longer. What is the ethical line between treating a disease and making a person stronger or taller or longer-lived? And what if such enhancements could be passed on to one's offspring? The latter question opens up the difficult issues inherent in the potential for germline gene therapy.

Germline gene therapy

In contrast to somatic cell gene therapy, which seeks to cure a single individual, **germline gene therapy** (sometimes called **inheritable genetic modification**, or **IGM**), seeks to eliminate "bad" genes both from the individual *and from that person's descendants*.

(A)   (B)

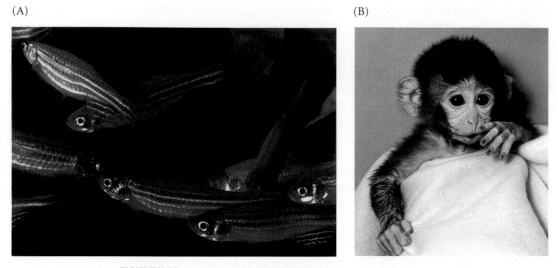

FIGURE 11.10   (A) Commercially marketed "GloFish" are zebrafish with a transgene for a protein that fluoresces (gives off light) in the dark. (B) A transgenic rhesus monkey named ANDi (backwards for "inserted DNA"). A transgene for green fluorescent protein is in each of his nuclei. (A courtesy of GloFish.com; B courtesy of G. Schatten.)

Germline gene therapy has not yet been attempted on humans. However, in laboratory animals it has been accomplished in two ways: (1) by modifying the parental germ cells or the fertilized egg such that a new genome is in every cell of the person's body and is therefore transferred to the next generation through the germ cells, or (2) by modifying embryonic stem cells so that the adult body contains a high percentage of cells derived from these genetically altered blastomeres.

Adding (or subtracting) genes from the germlines of laboratory mice has been a routine procedure for the past decade, and has vastly increased our knowledge of gene action (and interaction) in development. Outside the scientific community, commercial aquarium suppliers are already marketing transgenic zebrafish known as "GloFish." These fish contain a transgene for a fluorescent protein that makes them glow in the dark (Figure 11.10A).

There is little debate that such procedures could be done in humans should we desire to do so. Already, a transgenic rhesus monkey has been born that carries the gene for green fluorescent protein (GFP)* in each cell of his body (Figure 11.10B). The gene for GFP was inserted into a viral vec-

---

*Green fluorescent protein is a harmless substance produced by light-emitting jellyfish. First discovered in 1962, GFP is now a major tool used throughout the biological sciences, including wide use in the field of developmental biology. The *GFP* gene is a "marker" that can be linked to other genes, rendering the subject gene's product fluorescent (and thus readily visible).

tor that was injected into the space between the oocyte and the zona pellucida in a rhesus monkey egg. The vector inserted the *GFP* gene into the egg DNA, and the resulting offspring, a monkey called "ANDi" ("inserted DNA" backwards), has *GFP* genes in all of his cells (Chan et al. 2001).

In addition to the ethical and social arguments that will be discussed in Chapter 12, there are scientific issues that might mitigate against germline gene therapy. One of these concerns is the possibility that the introduction of a new gene might eliminate, or "knock out," a previously functioning gene. This would occur if the viral insertion vector were to splice its "genetic cargo" into a functioning gene rather than into a currently non-coding regions of DNA (Schröder et al. 2002). This has happened when scientists hoping to study the effects of one gene inadvertently "knocked out" another gene because the gene of interest fell inside the second gene's coding region.

The future of gene therapy

The past two decades have produced a tidal wave of new information about human genes and how they are activated. We now possess profound insights into how the body is constructed and how cells with the same genome (the same DNA) become different from one another. This knowledge has brought with it the potential to *change* the DNA of us and our descendants (to say nothing of manipulating the genomes of other organisms).

Whatever arguments may be brought against them, the technologies of gene therapy will most likely give us the ability to genetically modify our bodies (and the bodies of our children) within the decade. Some people look forward to the time when genetic diseases will be treated by gene therapy that repairs faulty genes in germ cells, thereby eliminating them from subsequent generations. Other people (in some cases, the same people) worry that this ability to manipulate genes will result in misguided attempts to enhance human physical or mental abilities. Such potential to knowingly affect the processes of evolution is a sobering concern to many, and raises questions that have no easy answers. As artist Stan Lee noted back in 1963, "With great power comes great responsibility."

> It is as if man had been suddenly appointed managing director of the biggest business of all, the business of evolution…
>
> JULIAN HUXLEY, "TRANSHUMANISM" (1957)

CHAPTER 12

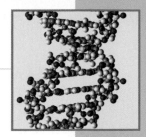

Should We Allow the Genetic Engineering of Human Beings?

T he sequencing of the human genome has opened up new vistas in medical research. As more and more genes and their protein products are identified, our knowledge of both genetic disease and the processes of human development becomes greater. With the application of this rapidly expanding knowledge to medical technology comes the prospect of a world free from debilitating genetic diseases. The bright, wished-for world most researchers envision is one where such human scourges as cancer, cystic fibrosis, and Parkinson's disease can be cured or even eradicated.

But many people see a "dark side" to gene therapy. The idea of deliberately altering a person's genes seems to some to be beyond the proper scope of medicine. And if the technology should be applied to nonmedical conditions—for example, altering the gene for human growth hormone merely to allow a child to be taller than his peers—a whole new spectrum of issues emerges.

As described in Chapter 11, the term "genetic engineering" can refer to either of two different forms of gene therapy, depending on the cell type being modified. **Somatic cell gene therapy** is essentially a medical treatment intended to target abnormally functioning genes in specific somatic (bodily) cells of a single patient. **Germline gene therapy** is intended to modify all of a person's cells, including the germ cells (sperm and egg). Not only is the person's genome altered by germline gene therapy, but the descendants of the person can inherit the altered genes. It is germline gene therapy that generates the most wide-ranging ethical questions.

Somatic Cell Gene Therapy

In the most common form of somatic cell gene therapy, a normally functioning gene is inserted (by means of a virus vector) into the patient's genome
(see Figure 11.9).The hope is that the inserted gene will be accepted by the
cell and translated into a normal gene product (a protein), thus relieving the
patient's suffering. Because the new genes are not inserted into the patient's
germ cells, no genome changes are passed on to offspring. Somatic cell gene
therapy is a medical intervention, like heart surgery or radiation therapy,
designed to cure or alleviate disease in a single individual.

Most scientists and medical professionals agree that treating a disease by
inserting a corrected gene into a patient is not ethically different from using
medicines to treat the disease. If sickle cell anemia or Huntington's disease
could be cured by administering effective drugs, doctors would certainly do
so; likewise, if they can be cured simply by adding a wild-type allele to a
patient's blood cells or neurons, few would quibble. Most of the concerns
voiced about somatic gene therapy are the same as those heard about any
cutting-edge medical advance: the safety of the procedures (see page 193)
and the equity of their availability (see page 170). In addition, however, gene
therapy raises the issue of medical enhancement (as opposed to medical
treatment).

The boundary between treatment (for a disease) and enhancement (for
cosmetic or athletic purposes) can be indistinct. For instance, is short stature
a disease? One could argue that taller people have a better chance of success than short people. Is infertility a disease? It prevents an individual
from reproducing, but does not usually cause harm in and of itself. Many
conditions not usually considered to be "diseases" are treated by the medical establishment, and some, such as baldness, are even covered by health
care plans.

Enhancement through gene therapy could conceivably become a huge
business. In the United States, the controversy over the use of dangerous,
illegal steroid drugs to enhance athletic performance shows how great the
desire can be for enhancement. Athletes are willing to risk severe potential
side effects and legal prosecution in order to gain the added strength and
power these drugs can offer. Gene therapy offers the possibility of enhancing strength and athletic performance without the risk of drugs, or even the
effort of exercise.

For instance, all vertebrates, including humans, possess a gene that
encodes the protein myostatin. This protein is a growth regulator that signals muscle cells when they have reached the proper size. Studies in mice
show that if the gene for myostatin is absent, the mice grow into muscular
rodents who are stronger and faster than their non-mutant littermates
(McPherron et al. 1997a). Individual muscles from *myostatin* mutant mice
weighed 2–3 times more than muscles taken from normal mice; the

increased muscle mass resulted from both an increased number of muscle fibers and an increased size of individual fibers. One of the discoverers of this gene commented that the myostatin deficient rodents "look like Schwarzenegger mice." News media printed articles about these mice, immediately linking them to what athletes might do if such therapy were available. Indeed, a human child has been found who is deficient in this gene. He has exceptional musculature and is much stronger than other boys of his age. Although he is healthy, physicians are watching him closely because the same gene is active in heart muscle, where such enlargement can be dangerous (Schuelke et al. 2004).

Whatever treatments or enhancements somatic cell gene therapy might provide, however, would be limited to a single treated individual. Germline gene therapy holds out the prospect of genetic alterations that could be passed along to future generations.

Germline Gene Therapy

In germline gene therapy, sometimes called inheritable genetic modification (IGM), the goal is to alter the genome at the germ cell (sperm or egg) level, so that the corrected or enhanced gene is transmitted to the person's offspring (see pages 195–197). Modification of germline genes has become a routine procedure using laboratory mice; it has contributed much of our knowledge about the actions and interactions of many vertebrate genes during development.

For UCLA professor Gregory Stock, a proponent of genetic engineering, the question is not *if*, but *when* we will be able to engineer the human germline. However, the discourse surrounding germline engineering is full of controversy. There are strong advocates of furthering the fledgling technology, but many also believe that germline engineering research should be strictly regulated or banned. Given that germline gene therapy might be able to eradicate inherited genetic diseases and enable us to expand our genetic repertoire, why should anyone be against it?

> [O]nce a relatively inexpensive technology becomes feasible in thousands of laboratories around the world and a sizable fraction of the population sees it as beneficial, it will be used.
>
> GREGORY STOCK, *REDESIGNING HUMANS* (2002)

In fact, many people question whether the therapy is even needed for medical purposes. The American Association for the Advancement of Science (AAAS) has indeed identified a few specific instances where the germline modification approach could be used to prevent parents from transmitting defective genes to their offspring. But several alternative procedures—including prenatal genetic diagnosis, gamete donation, embryo selection, and adoption—are currently available and would not evoke the issues involved in germline manipulation (Chapman and Frankel 2003), so there might not be a great medical need to do it. Other arguments against IGM range from the pragmatic to the emotional.

"It isn't safe"

One safety issue that applies to both somatic and germline therapy concerns the way in which the corrected DNA is inserted into the target cells. As described in Chapter 11, these methods often make use of viral vectors, and such vectors can trigger massive, systemic immune responses (see page 193).

Another argument used against IGM is that when altered genes are inserted into the genome, they may disrupt presently functional genes and cause mutations. This has certainly been encountered in laboratory mice (Gilbert 2003). In one case, the disruption of a single gene resulted in mice who were born without eyes, semicircular ear canals, or a sense of smell (Griffith et al. 1999). Another transgenic strain of mice developed normally but began to develop tumors at a very high rate early in their adult lives due to a malfunction in the regulatory mechanisms of the inserted gene (Leder et al. 1986).

Proponents of IGM point to advances in targeted gene insertion as an indication that stumbling blocks may yet be overcome (Waters 2002; Belteki 2003). However, in the cases of inheritable manipulations to the germline, some effects may take several generations to manifest themselves—and any mistakes made will be permanent. IGM is not a drug that can be discontinued if the side effects are disastrous (Newman 2003). Mario Capecchi has considered this problem, and has devised a system by which germline engineering could become, in effect, "whole body somatic therapy" (Ponturo 1998). Capecchi's proposed technique would use artificial chromosomes to add genes that could be induced to lose their centromeres by a signal from an exogenous (i.e., coming from outside the body) factor. Once the signal was administered, such chromosomes would not survive meiosis or mitosis and would not be passed on to the next generation.

"We are playing God"

This argument is heard whenever a new technology is invented or proposed. It was used against Benjamin Franklin's lightning rod and it was used against smallpox vaccination. This is akin to the emotional argument that says we should not interfere with nature (White 1960). But medical interventions are never undertaken with the view that nature is benign, and are by their very existence interfering with nature. One can—and some people do—accuse heart surgeons who perform bypass surgery or neurosurgeons who remove brain tumors of "playing God." Most of us, however, are glad that they do so. Indeed, in some religions, "playing God" by healing is considered a most worthy endeavor (see pages 171–172).

"We do not know what such genetic technology will be used for"

If lethal genetic diseases such as Lesch-Nyhan syndrome or Huntington's disease can be screened for by preimplantation genetics, germline genetic engineering becomes a very high-tech solution for a problem that has a relatively low-tech cure. So what else might the technology be used for? One possibility is that it could be used for phenotype enhancement.

Modern plastic surgery has allowed thousands of people to live better lives. The ability to restore facial or limb structure and function to those who have lost it due to trauma or because of genetic malformations is one of the highest achievements of modern surgical technology. In addition to such cases, however, millions of people make use of these same surgical techniques in order to make themselves more attractive. Similarly, medical advances used to fight prostate tumors can also be used to prevent male baldness (see page 195). It is reasonable to expect, then, that genetic technologies designed to fight diseases will also be used for purposes of enhancement. We know there are genes that affect height and muscle mass, so we could conceivably make our offspring taller and stronger. If genes involving intelligence were found, those who could afford this procedure might enhance themselves in the hopes of producing highly intelligent offspring.

Whereas somatic cell gene therapy is like any other medical procedure in providing value for the patient, IGM differs in providing value for the patient's offspring. Jeremy Rifkin (1998) voices his concern that "those families who can afford to program 'superior' genetic traits into their fetuses at conception could assure their offspring an even greater biological advantage, and thus a social and economic advantage as well." Lee Silver (1998) envisions a world where, due to such economic inequality, the genetic haves and the genetic have nots are far apart in their abilities: genetic engineering would convert economic differences into inherited biological differences. At the moment, there are no restrictions on what such therapy could be used for. Unless legislation can define what is allowable and what is not allowable, this remains an important critique of the technology.

"Do we really know which traits to enhance or get rid of?"

Genetic engineering assumes that we know which traits are good and which are bad. However, what is good in one environment might be harmful in another. The genetic mutation that causes sickle cell anemia is deleterious when homozygous (that is, when it is inherited from both parents)—but when only one copy of the gene has the mutation, the overall effect may be advantageous in certain environments since it may offer some protection against the parasitic disease malaria. A mutation in a gene for a certain molecule on lymphocytes may normally be a bad thing; but this same mutation may possibly offer protection against HIV, the virus that causes AIDS. Similarly, the same mutation that predisposes people in Western populations to allergic reactions and asthma is advantageous in areas of the world where certain parasites are a major health problem (Rockman et al. 2003).

If we were to know which genes or groups of genes produce aggressive or docile phenotypes should we change them? Are certain alleles or combinations of alleles deleterious in some situations but in others predispose towards acts of genius? We just don't know. There exists a fear that the traits people choose for their children will not be healthy in the long run. Some question the consequences if a trait chosen in one generation falls out of fashion in the next or becomes particularly ill-suited to a change in the environment (Council for Responsible Genetics 2001).

"Do we even know the functions of the genes that might be changed?"

It is one thing to look at genes for proteins that are the end-products of development—hemoglobin or insulin, for example. These genes probably have a single function. But genes that act during development often have many functions, a condition called **pleiotropy**. Expression of the *BMP4* gene, for example, can induce bone growth in some tissues but induces apoptosis (the death of cells) in a different set of developing tissues. In ectodermal tissues, the same *BMP4* gene product can induce cells to differentiate into epidermis (skin) instead of nerves. We are constantly discovering that genes are not "for" a particular function; rather they are "used in" a particular function. If we alter a gene in order to affect one function, we may well find that it also disrupts another function.

"Do parents have the right to make decisions about their children's genotypes?"

In the normal scheme of things, there is a great deal of chance involved in which traits a child will inherit from its parents. But what if the child's genes were "designed" and paid for by the parents? The parents would directly control the qualities of their offspring. If inheritance of certain traits were a certainty, the individuality of the child could be affected. If parents were to select genes for height and body musculature, they might then pressure their child to succeed at sports, regardless of whether the child *wants* to play the game. The entire notion of individual personhood is called into question. These questions are similar to those raised by the issue of reproductive cloning (see Chapter 8).

A range of critics believe that germline genetic engineering could convert a child into a commercial product with expected parameters of normalcy and function. Such critics maintain that one might get "fads" in children—one generation preferring a certain hair color, height, or organ endowment in its children—and the standards for what is genetically desirable would likely be those of the society's economically and politically dominant groups. People who fell short of some technically achievable ideal would be seen as "damaged goods," increasing prejudices and discrimination (see "Toward GATTACA," page 105).

Disability rights advocates are critical of germline engineering technology because they fear that a social objective of establishing the "perfect" human might lessen society's value on care and respect. In addition, the loss of care and respect for the less fortunate would leave disabled people as pitied mistakes, born with genetic diseases that could have been corrected.

It should be noted that not all genetic diseases are foreseeable. The hyperbolic notion of a world (or a country) free from genetic disease is not going to happen. For instance, the most common form of dwarfism—achondroplasia—is caused by a dominant genetic mutation that causes cartilage in the arms and legs to stop growing too soon. About 7 out of every 8 cases of achondroplasia are not inherited but are due to new mutations, mostly carried on the sperm (Tiemann-Boege et al. 2002). These are random events that cannot be predicted. (And, as noted in Chapter 13, many of these little people do not find their condition to be a "disease.")

Most scientists believe that *every* human being is a heterozygous ("recessive") carrier for several harmful genes. In other words, every one of us carries one "bad" copy of at least a few genes. This is why marrying close relatives is a dangerous enterprise. In most instances, an unrelated couple will not both carry bad copies of the *same* gene. However, if close relatives marry, the odds rise that they will carry the same mutant recessive genes and that their offspring will be adversely affected.

"Genetic engineering may lead to eugenics"

Eugenics means "well born." Rising from a movement beginning in the late nineteenth century, eugenics is a program that advocates breeding better humans (like breeding better crops and livestock). It was a major movement until the end of World War II. Eugenics attempted to make the human race more uniform and healthy. While the historical movement was based on unsound biological principles, eugenic goals might now be achieved through biotechnology. But such engineering of the genome might have consequences in reducing biological diversity (Allen 2001; Duster 2003).

Communities of color have historically suffered from the racist social applications of eugenics theories, and they are concerned that germline engineering offers another opportunity for racism to manifest, veiled as science (see Selden 1999). Although most scientists involved in germline engineering have no explicit racist agenda, civil rights advocates have found it disconcerting that David Duke, former National Director of the Ku Klux

The Eugenics Movement

The Englishman Francis Galton coined the word *eugenics* ("well born") in 1883 to express his idea for the programmed improvement of the human species. Eugenic goals included the elimination of genetic diseases and the enhancement of desirable traits. In 1908, Galton organized the Eugenics Society in London; similar societies were established around the same time in the United States and Germany. Their goals were to investigate human heredity and to carry out social action programs aimed at preventing the spread of "disadvantageous" traits (negative eugenics) while encouraging the transmission of "advantageous" ones (positive eugenics).

Although some members of these groups were motivated by high ideals and altruism, others were preoccupied with issues of race, class, privilege, and with retaining their own positions of influence in what they perceived to be a degenerating society. With its basis on classifications of "superior" and "inferior," eugenics had appeal for both idealists and cranks. Galton himself (who, somewhat ironically, never had offspring of his own) advocated only voluntary actions, but his aspirations were heavily tinged with nationalism. He noted that a "high human breed" was especially important for the English because they had colonized the world and planted the seeds of "future millions" of the human race (Mange and Mange 1999).

> Any farmer would promptly predict the fate of a herd of cattle in which the scrub stock was allowed to breed faster than the pedigreed stock. Yet there is no doubt that in civilized countries large families are the rule among the undesirable elements and the exception in the best stock.
>
> ELLIOT R. DOWNING,
> *ELEMENTARY EUGENICS* (1928)

By the 1920s, the eugenics program had become particularly strong in the United States. Fueled by fear of "racial degeneration" from the large families of the waves of impoverished immigrants then arriving from central and southern Europe, American eugenicists influenced many aspects of social policy (Eugenics Archive 2005). Among their "victories" were the enactment of the discriminatory Immigration Restriction Act of 1924, and the involuntary sterilization of thousands of individuals deemed to be "unfit" (Ludmerer 1972; Kevles 1995; Selden 1999). In the 1930s, many American eugenicists applauded the sterilization measures carried out by Nazi Germany in their glorification of Aryan elitism (Müller-Hill, 1988; Allen 1996). However, when the results of Hitler's endorsement of eugenic notions culminated in the horrors of the Holocaust, most biologists left the eugenics movement.

Eugenicists had faulty knowledge of genetics and a flawed and biased view of ethnic traits as genetic rather than environmental or cultural. They falsified data in order to further their own ends, promulgating a frightening view of a world overrun by imbeciles and criminals that would result if their mandates were ignored. However, as the evolutionary geneticist Theodosius Dobzhansky observed, the world's problems are not caused by the genetically impaired, but by people with superb genes who use their talents for amoral ends (see Dobzhansky 1962, 1976).

Klan, heartily supports inheritable genetic modification development. The case for genetic engineering hasn't been helped by those scientists who attempt to promote it by claiming it will "cure" homosexuality, criminality, and homelessness. In the past, the concept was that eugenics would arise from government policy (as it did in Nazi Germany, with historically horrifying results). However, such policies could also come from social pressure and economics (Paul 1996).

The Public Policy Debate

There is a conundrum inherent in regulating cutting-edge scientific and medical technology. While it is illogical to create restrictive legislation to cover an as-yet unknown technology (especially one with the potential ability to benefit society), once a new technology is in use and embedded in society, it may be too late to create effective and comprehensive policy. Consequently, public policy for germline engineering has been difficult to construct and different nations have approached it differently.

Germline engineering is a particularly sensitive subject for policy makers not only because it allows science to tinker with life and death, but also because changes to the germline affect future generations, with the possibility of affecting the trajectory of human evolution. In the United States, there are many voices of caution and many advocates of unrestricted research, both of whom are trying to influence U.S. policy in this time of flux. Ideally, policy makers would weigh the various considerations of the public interest and decide upon a thoughtful and comprehensive public policy regarding germline engineering. In part because of the conflicting ethical, scientific, and policy advice coming from all sides, the U.S. government has not yet created any binding policy.

A brief political history

Scientific proposals asking for funding for experiments in manipulating the human genome were first penned in the early 1980s (see Maienschein 2005; Gonçalves 2005). These proposals were very controversial; only a minority of scientists favored pursuing the research, and public sentiment was generally absent or negative. In 1983, a letter organized by anti-biotechnology activist Jeremy Rifkin (see Rifkin 1998; Hayes 2001) and signed by 53 religious leaders declared that genetic engineering of the human germline "represents a fundamental threat to the preservation of the human species as we know it." In 1994, the European

> Genetic engineering of the human germline represents a fundamental threat to the preservation of the human species as we know it, and should be opposed with the same courage and conviction as we now oppose the threat of nuclear extinction.
>
> THEOLOGICAL LETTER TO THE
> UNITED STATES CONGRESS (1983)

Scientific Commission Chairperson, Noelle Lenoir, said germline therapy was far too risky, and that "such experiments should not be conducted on humans" (Hayes 2001). Germline enhancement was new, scary, and unpopular. John Gearhart of Johns Hopkins University summed up the climate of the 1980s, saying that "no one from the conventional scientific world would try germline therapy, for fear of ostracism and out of ethical sensibility" (Airwin 1997).

Even amid the negative public sentiment, the U.S. National Institutes of Health (NIH, the primary government agency funding medical research), received proposals to fund human somatic and germline engineering experiments. The government agency that received the task of systematically deciding whether such proposals should be funded was a small branch of the NIH called the Recombinant DNA Advisory Council (RAC). Any recombinant DNA research conducted at institutions receiving federal funds, be they private or public, are regulated by the RAC. In 1985, the RAC decided that it "[would not then] entertain proposals for germ line alterations but will consider proposals involving somatic cell gene transfer as long as the proposals meet the established criteria" (Recombinant DNA Advisory Committee 1985).

Advocates of germline engineering saw this decision as the first step toward human laboratory experimentation, while opponents hoped that the NIH had drawn their final ethical line. Meanwhile, research continued on non-inheritable somatic gene therapy, quietly paving the way to engage in germline engineering clinic trials. The NIH will still not entertain proposals on human germline engineering, but social and political changes have occurred that make a future shift in NIH policy possible and even likely.

By the end of the millennium, the opinion of the scientific community was generally more amenable to germline engineering than it was two decades earlier. In 1998, a symposium entitled "Engineering the Human Germline" was held at the University of California, Los Angeles; this meeting may have been a watershed moment for pro-germline engineering forces. The UCLA symposium—whose principal organizers were academics interested in pursuing germline research—gathered together many germline engineering advocates and endowed their individual voices with the strength of institutional academia and scientific consensus. The conference was well attended, well respected, and well documented; over 1,000 people came, and it received major media coverage, including front-page stories in the *New York Times* and *Washington Post*. Although there was some negative response, the majority of published opinions were more positive than negative. According to the *New York Times*, "the scientists, leaders in the fields, were meeting on their own, with no government or other mandate to issue guidelines or regulations and, in fact, no wish to restrict their work" (Kolata 1998). In other words, the UCLA symposium lent a sense of inevitability to the progress of germline engineering research.

In addition to creating a large and positive public sentiment surrounding germline engineering, the UCLA symposium produced a test case that would allow the NIH to agree to the possibility of germline engineering without explicitly approving the technique. W. French Anderson, an experimental scientist, proposed clinical trials involving the transfer of somatic genes to human fetuses. Although no germline genes would be experimentally manipulated, Anderson's proposal acknowledged that there would be a "relatively high" potential for "inadvertent gene transfer to the germline." The NIH has not yet funded the proposal, but Anderson still hopes to receive its approval; if he does, the door to federally funded testing on the human germline will be knocked ajar.

The NIH received over 70 pages of public comments about Anderson's proposal, the overwhelming majority of which opposed germline gene therapy (Begley 1998). The response indicates that public opinion is not as universally positive as the UCLA symposium sponsors might wish; germline engineering is still a "hot" and undecided issue.

Legally, there is no ban in the United States on germline engineering research, so a rejection from the NIH only means that Anderson and others must find private sources to fund their research. Indeed, private sources have funded experiments similar to those proposed by Anderson. In 1996, doctors at an infertility clinic in New Jersey started using a crude form of germline manipulation called "cytoplasmic transfer" to help their patients achieve fertility. The doctors, employees of the Institute for Reproductive Medicine and Science in Livingston, New Jersey, helped their clients conceive by slightly altering the cytoplasm of the infertile woman's egg. Essentially, the doctors took a client's fertilized egg and added 5% of the cytoplasm (including mitochondria, which have their own genes) from a donor egg to replace any malfunctioning units in the client's egg. Between 1996 and 2001, a reported 16 babies with the genetic makeup of three parents (i.e., the client mother and father, and the female cytoplasm donor) were born as a result of this technique (Mangles 2001; Meek 2001).

> All of the reasons people have given for saying [germline gene therapy] is wrong are either irrational or religious-based. Some people say we should not go against nature, but that's illogical because every time we cure a disease we go against nature.
>
> LEE SILVER (1998)

The Institute for Reproductive Medicine and Science is a private facility and its experiments are not covered by the ban on federally funded embryo and germline research. Therefore, although some might consider their actions unethical, they were not unlawful. Furthermore, their goal was not to create a "designer" baby, but to help women conceive. Nevertheless, this new germline manipulation made some legislators uncomfortable and the government took action. The U.S. Food and Drug Administration (FDA) declared the clinic's fertilization technique to be a procedure under their jurisdiction and ordered the clinic to stop using the procedure (Brave 2001). In addition, the genetics-focused ethics panel of the National Institutes of Health and the White House's Bioethics Commission decided to look at developing guidelines for broader oversight of human genetic experiments (Mangles 2001).

These steps may have temporarily resolved the specific situation in one New Jersey clinic, but larger policy issues have been left ambiguous. In the New Jersey reproductive clinic case, the FDA's authority might seem tenuous at best. The FDA is a part of the executive branch of the federal government; its mandate is to protect the public health, mainly by the examination and monitoring of consumer products. According to their website, "the FDA's mission is to promote and protect the public health by helping safe and effective products reach the market in a timely way, and monitoring products for continued safety after they are in use. Our work is a blending of law and science aimed at protecting consumers" (FDA 2004).

But the FDA also has the authority to regulate *research* into products that have an intended human use and if the products "move in interstate commerce." Therefore, if the products being researched are intended for human use—and many biotechnology applications are—the FDA claims the right to regulate the research (National Bioethics Advisory Commission 1999)

Recent legal precedents support an extension of the FDA into germline engineering research regulation. Court decisions granted the FDA the power to regulate products that substantially affect interstate commerce, even if the products do not move in interstate commerce. It would seem that all research on germline enhancement would affect interstate commerce, since no research is likely to remain in a single state. Thus, the FDA could become an influential regulator of all public and privately funded germline enhancement research. Additionally, because the FDA requires any procedures submitted to it involving recombinant DNA to also pass RAC review, involving the FDA in private sector research effectively enables the RAC to supervise and regulate private research (FDA 2000).

Current government positions on germline engineering

As of 2005, the field of germline engineering is relatively small, and it is dominated by a few key individuals and institutions. In the United States, it is possible that the FDA and RAC will continue to serve as a watchdog on the private sector in the absence of actual legislation, or perhaps legislation will supplant them.

There are a several public policy strategies open to the U.S. The default state is the libertarian position of leaving such decisions to the market place (that is, no regulation). The other extreme would be to create laws that ban outright all germline engineering research or application.

Between these two extremes are positions advocating governmental regulation. In other words, germline research and applications might be found warranted in some cases but not in others. The government could also act by forbidding to grant patents resulting from genetic engineering technology or information—a decision that would decrease incentives in the private sector to pursue new technologies related to the human germline. Lastly, the public could decide not to use genetically engineered products, thus lessening the demand for such services.

Government officials do not shape policy alone; instead, that process is informed and influenced by scientists, non-government organizations like bioethics councils, the business community, religious groups and public opinion. The President's Council on Bioethics, the Center for Genetic and Society, Genewatch, The Council for Responsible Genetics, and the American Academy for the Advancement of Science all contribute to policy decisions.

Other countries have passed legislation banning germline engineering. Australia, Austria, Costa Rica, Denmark, France, Germany, Hungary, India, Israel, Japan, Norway, Peru, Spain, Sweden, Trinidad and Tobago, and the United Kingdom have all passed laws or regulations that somehow proscribe or limit human germline engineering. International groups have also discussed genetic modifications and hope to impede the speed of science by interjecting ethical, environmental, and religious considerations into the decision-making process.

Currently, the Council of Europe's 1997 Convention on Human Rights and Biomedicine is the most comprehensive and authoritative international agreement. The Convention bans inheritable genetic modification and bans or regulates many other human genetic technologies. Another non-enforceable document was created in 1997 by UNESCO, the United Nations Educational, Social, and Cultural Organization. UNESCO adopted a non-binding Universal Declaration on the Human Genome and Human Rights, part of which told UNESCO's International Bioethics Committee to study "practices that could be contrary to human dignity, such as germ-line interventions" (UNESCO 2003). The Universal Declaration was signed by 186 nations. Although it would be ideal to have an international body capable of regulating scientific research and the advances in biotechnology, UNESCO does not have that power or authority. Ultimately, the United States (or any other nation) can create its own legislation and can choose to ignore UNESCO's recommendations.

The United Nations' World Health Organization is a somewhat more powerful international governing body, and mandates passed down by WHO are often enforced worldwide. In 1999, a major WHO study on genetics, cloning, and biotechnology included a public health component that explicitly called for a global ban on inheritable genetic modification. However, these guidelines were never published in final form. More recently, WHO established a different advisory committee to enable policy decisions to be made about genetic technologies; its final advice is yet to be determined.

In sum, the landscape of international operational policy and U.S. domestic policy are currently in a state of flux. The United States has no mechanism in place to stop the private sector from engaging in germline research or practices. Such a mechanism would require an act of Congress and a slow legislative process loaded with ethical decisions. Leaving germline engineering free to develop in an unregulated market does not allow the public moral interest to affect the trajectory of germline engineering. Ultimately, the United States will have to make a legislative decision, or else allow the private sector to dictate the next steps.

UNIT 7

New Perspectives on Old Issues

I have studied philosophy, law, and medicine, and worst of all theology;
and here I am, for all my lore, the wretched fool I was before.

GOETHE, *FAUST* (1808)

CHAPTER 13

What Is "Normal"?

> My life has been happy because I have had wonderful friends
> and plenty of interesting work to do. I seldom think about my
> limitations, and they never make me sad. Perhaps there is just
> a touch of yearning at times, but it is vague, like a breeze
> among flowers. The wind passes, and the flowers are content.
>
> Helen Keller (1927)

"Is my child normal?" is usually one of the first concerns of new parents. Indeed, genetic screening and prenatal diagnosis are touted as ways of ensuring that one's children will be normal. But what does "normal" mean? It is a word that is often highly charged with subjectivity and prejudice, although attempts have been made to strip away the value judgments in order to use the term in an unambiguous and non-pejorative manner. The concept of "normalcy" or "normality" is widespread both in medicine and in everyday thought, and what we are usually trying to articulate is the sense of a normal phenotype.

The term "phenotype" is not as familiar to most people as the word "normal," but it is the easier of the two terms to define. The prominent evolutionary biologist Theodosius Dobzhansky characterized the phenotype as "the total of everything that can be observed or inferred about an individual" (Dobzhansky 1962). The phenotype encompasses both external appearance and internal anatomy and physiology, as well as a person's behavioral interactions with their family and with society as a whole. A person's phenotype is the result of their genotype (the total output of all their genes) and its interaction with the environment they encounter as they develop (see Chapter 14).

Defining Normal

According to Lennart Nordenfelt, the author of numerous books and papers on human ability, quality of life, and health care, the concept of normalcy can be defined according to two different standards: in terms of a statistical average (which is an objective mathematical calculation), or in terms of some subjective "ideal" performance. The statistical view attempts to define "normal" while circumventing value judgments. On the other hand, inherent in the concept of an ideal performance is the assumption that normal is superior to abnormal. The distinction between these classifications often blurs, and even when purely descriptive accounts are attempted, superiority of one phenotype over another is often inadvertently implied (Nordenfelt 1995).

Edmond Murphy (1972) lists seven overlapping uses of the word "normal" in the medical literature:

1. It can refer to a statistically normal distribution, as in "American men are normally between 5 and 7 feet tall." Murphy suggests using the word "Gaussian" rather than "normal" in this context. (Gaussian, after the mathematician Karl Friedrich Gauss, refers to a statistical probability distribution around a most common value. It is based on a mathematical function that describes the familiar "bell curve.")
2. "Normal" may refer to the"most representative" or "most commonly encountered," in which the term "usual" or "general" might be more accurate, as in the phrase "humans normally have two eyes."
3. In genetics, "normal" is often used to refer to the "wild type," or the most standard genotype/phenotype found in natural settings (and thus the reference point against which other individuals of the same species are judged).
4. Clinical medicine often defines "normal" as "carrying no penalty" which might be translated as "innocuous" or "harmless," usually when referring to physiological function.
5. In sociology and politics, "normal" is often used instead of "conventional" or "unquestioned." Thus we normally assume it is better to have sight than to be blind.
6. In aesthetics, "normal" can mean "the most perfect of its class" or "ideal." This might better be termed "the standard."
7. "Normal" is often considered to be that which is distinguished from the "pathological" or "diseased."

As embodied in the final definition, one of the battles being waged concerns exactly what constitutes a "disease"—that is, what is normal and what is abnormal. This question brings up the related issue of a process that critics call "medicalization," whereby some trait that has always been thought of as normal comes to be viewed as a medical condition. For instance, Leonore Tiefer describes how the pharmaceutical industry has attempted to convert the personality trait we characterize as "shyness" from a normal human condition into an "affective disorder" that needs medical diagnosis and treatment, and how areas of women's health such as age-related loss of sexual desire are now seen to require "medical surveillance to assess and monitor deviations from alleged health norms" (Tiefer 2004). In these examples, we see how a behavior that might be on one end of a "normal" (Gaussian) curve has been separated from the "normal" range and classified as a qualitatively different category. We also see that concepts of normalcy can change with time and social use.

This chapter is primarily concerned with what society perceives as the normal human condition in terms of physical function. People vary widely in their ability to function, and at least some of this variation is based in their genes.

Normal Variation among Individuals

In the 1950s and 1960s, Theodosius Dobzhansky's work with genetics showed that there is such great genetic diversity among individuals in a population that no one state can be considered "best." Geneticists sometimes use the criterion that if a genetic variant is present in at least 5 percent of a population, it represents a normal (i.e., expected) variation. Thus, a gene variant whose prevalence is less than 5 percent is referred to as a mutation, while any variant present in at least 5 of every 100 individuals is part of normal, expected variation (and is called a polymorphism). This somewhat simplistic criteria is perhaps more easily applied to other species than to humans, but it does allow a point of reference to discussing normal phenotypes.

In the years after Darwin, some thinkers based their concepts of normality on the survival of a species and natural selection. According to this way of thinking, known as adaptationism, the trait best suited to the environment is the biologically superior trait and becomes the most common characteristic, or the optimum. Deviations from this optimum were believed to be biologically inferior (Vacha 1985). A similar theory was supported by Alfred Russel Wallace and several other evolutionary biologists but was strongly opposed by Darwin himself (Gould and Lewontin 1994).

Adaptationist theory was predicated on the assumption that evolution is driven by a force whose goal is perfection. If this force is assumed to exist, we can indeed reach the conclusion that those traits selected for most often are inherently better than other traits, at least in a given environment. But, as Darwin pointed out, evolution is *not* goal-directed. Natural selection is a random process that *requires* the presence of variation within a population. Without variation, a population cannot adapt and is vulnerable to extinction if the environment changes. Variation itself is normal. A normal population will encompass a great deal of genetic variation, and most individuals within a population will deviate from the average (Davis 1995). Moreover, over time, the proportion of the population with one genetic variant or another will drift at random. This random genetic drift plays an important role in shaping traits over the course of evolution (Futuyma 2005).

When considering modern human communities, we also have to take into account those aspects where natural selection exerts much less pressure than in natural communities of other species. A simple condition such as poor eyesight—easily correctable with eyeglasses in modern human societies—might have been a severe handicap for prehistoric humans who depended on their hunting skill. Conversely, a trait that may have once helped individuals may now be a detriment.

For example, geneticists are studying Pima Indian populations in the southwestern United States, among whom there is a much higher incidence of obesity, heart failure, and adult (type 2) diabetes than is average for the United States (an average that is already higher than most of the rest of the world's). It appears that over many centuries of living in a harsh desert environment, where food was often scarce, the ancestors of today's Pima may have been subjected to intense natural selection for those gene variants whose protein products maximized the efficiency with which individuals converted food into energy and energy reserves (i.e., fat). Today's Pima Indians have access to the plentiful food supply common in the United States, including many foods high in sugar and fat. Their inherited digestive genes are still more efficient than most people's at converting food into energy and fat—which in their current environment is not a good thing for their overall health (see http://diabetes.niddk.nih.gov/dm/pubs/pima/genetic/genetic.htm).

Norms Determined by Functional Ability

In 1945, an article in a Yale University medical journal argued that normal must be defined as "that which functions in accordance with its design" (King 1945). In 1964, a standard textbook of medical pathology, *Principles of Pathology* (Hopps 1964), asserted that there is a single normal standard for function, and anything that differs from that standard is abnormal.

In modern society we tend to categorize those with significantly less functional parts as "disabled," a category that relies heavily on the *functionality* models of normality and classifies disabled people as those who depend on others for help with basic needs. However, physical aids, such as wheelchairs and hearing aids, can improve function beyond what is accepted as normal. For example, although the world record for a marathon is 45 minutes faster for a wheelchair user than the fastest runner, wheelchairs are often shunned as a less efficient, cumbersome means of movement that should be avoided if possible. It is acknowledged in many physical rehabilitation programs that cosmetic normalcy is a standard that should be sought, even at the expense of functionality (Amundson 2005).

Interestingly, there are times when the definitions of normalcy determined by functionality contradict those determined by statistics. For example, the Pima are far from the only Americans who suffer from health-impacting obesity. Roughly 25 percent of U.S. adults are medically obese (CDC 2005); thus, if we apply the "5 percent rule," obesity in the American population is certainly a normal condition, even though it can be a functionally deficient (i.e., unhealthy) one.

A Modern Approach to Disability

Today we are on the threshold of being able to understand each gene's product and its role in the functioning of "normal" people (Wertz 2002). Prenatal genetic screening allows parents a glimpse into their child's potential future before the child is born. Given this information, parents can decide whether they think their child's life will be worth living, terminating the pregnancy if they disapprove of the fetus's genetic make-up. But how does one define which traits are acceptable to carry to term and which are not? Will disability become a class-based phenomenon in which only the wealthy and middle-class subsections of the population will be able to afford screening?

According to a 37-country survey conducted by Dorothy Wertz and Jon Fletcher of the Shriver Center, there is currently little consensus regarding the ethics of prenatal selection. When questioned by the Shriver Center, 85 percent of U.S. geneticists supported abortion of fetuses with Down's syndrome, 92 percent would advise the termination of fetuses with severe open spina bifida (exposed lower spinal cord), and 56 percent would warn against bringing a fetus with achondroplasic dwarfism to term. The public tended to be more sympathetic to genetic conditions and only gave substantial support for selective abortion if the fetus were so mentally retarded that it would never be able to understand language (Wertz 2002).

Prenatal diagnosis has been a blessing to many families. Indeed, after experiencing the trauma of watching their babies die of genetic diseases, some families had vowed never to have another child. But prenatal genetic diagnosis allows them to know in advance that a child would not succumb to the same diseases, thus allowing such couples to have healthy children. Nobel laureate James D. Watson has exhorted geneticists that "we place most of our hopes for genetics on the use of antenatal diagnostic procedures, which increasingly will let us know whether a fetus is carrying a mutant gene that will seriously proscribe its eventual development into a functional human being. By terminating such pregnancies, the threat of horrific disease genes contributing to blight many family's hopes for future success can be erased" (Watson 1996).

Watson's way of thinking worries some who are themselves disabled or who lobby on behalf of those who are. Adrienne Asch (1999, 2003) has criticized prenatal testing programs for being misguided and based on mistaken premises concerning disability. First, she notes, prenatal testing for genetic diseases reinforces the medical model that it is the disability itself, and not the social response to the disability, that is the problem to be solved. Asch presents evidence that it is the social discrimination that is far worse than the physical disease in many cases. Second, she claims that the prospective parents are being told that having a child with a disability will diminish their parental joy. Here, parents become fixated on the range of opportunities (that has become limited) rather than on the possibilities of rewarding choices within that range. And third, prospective parents are led to believe that disabled people cannot have a meaningful and happy existence. She believes that parental expectations of what is good or valuable are far too limited. One might be blind, or confined to a wheelchair, or unable to hear, but this does not preclude a rewarding career, love, or deep relationships with friends.

The Shriver Center study conclusions indicate that social stigmas, supported by "scientific objectivity," exert great pressure on individuals to conform to what is deemed phenotypically normal. And because individuals are influenced by the society of which they are members, helping individuals to conform is consistent with helping society conform, and only serves to strengthen the stigmas. Finally, the report voices concern that as prenatal testing and selective abortion become more common and available to the wealthy, the shame of disability will become increasingly entangled in issues of class (Wertz 2002).

For groups of people who have significant decreases in specific functions, the concept of "normality" has often polarized them into "integrationists" and "separatists." Integrationists applaud attempts to "mainstream" their group into the larger society. Separatists feel that such attempts to enforce people to become "normal" according to standard definitions will put members of their community at a disadvantage. Rather, the separatists would prefer to set their own norms according to their own social conventions.* We discuss here examples of two such communities where this is happening in different ways: the deaf community and the intersex community.

Deafness

In some circumstances, deaf individuals have "normalized" their biological deficiency by forming their own norms, which they claim to be as valid

---

*A historical parallel might be made with the American civil rights movement of the 1960s and the polarization between the followers of Martin Luther King and those of Malcolm X.

as those of the larger society. Since the early eighteenth century, when most of the deaf found themselves isolated in a hearing world, deaf members of society have come together, developed a language, and established a way of life. To some within the deaf community, the deaf way of life is not only completely normal, but is also desirable and worth defending from those who wish to destroy it—those who view deafness as a disease and a disability that needs to be "cured."

The history of the deaf culture begins in the early eighteenth century, when hearing people discovered through reading and writing that the deaf persons are as adept as the hearing at processing language. The Abbé de l'Epée, founded a public school for the deaf in the late eighteenth century, and in 1771 he began displaying his students' accomplishments to crowds in Paris. The deaf students responded to thought-provoking question in written English, French, Latin, German, Italian, and Spanish (Davis 1995).

It was around this time that sign language began to develop from written language, and deaf people started to come together to form a society. The hearing world of the nineteenth century viewed this deaf community as its own ethnicity or race (Davis 1995) and ultimately most came to consider them as inferior citizens. Alexander Graham Bell, inventor of the telephone, was a passionate eugenicist who thought that the deaf would contribute to "the production of a defective race of human beings, [which] would be a great calamity in the world." He believed that sign language should be forbidden, education through sign language should be abolished, and that deaf people should not be allowed to teach other deaf students (Bell 1869).

Today such organizations as the National Association for the Deaf (NAD) exist to support the hearing-impaired and their culture. However, to many, deafness is a disease that needs a cure, and modern technologies designed to aid the deaf have led to controversy. The integrationists view such technologies as a way of joining mainstream society, while the separatists see them as taking away their own society, making deaf people into second-class citizens of the larger culture. This struggle within the deaf community was dramatized in the play (and later the movie) *Children of a Lesser God.*

Perhaps the most controversial of all the technological advancements has been the cochlear implant. This apparatus consists of an external microphone connected to a device that is surgically implanted in the patient's cochlea. The microphone transmits speech and other sounds to the implant, which converts the signal into an electrical impulse that the auditory nerve can conduct to the brain. The NAD asserts that these devices do not facilitate language acquisition or educational success in deaf students; however, the National Institutes of Health found that the value of giving implants to children under the age of two is much higher than implantation after that time. After the age of two, children have passed the critical period for audible language acquisition (Finn 1998).

Although cochlear implants could ideally allow deaf people to be included in both the hearing and the deaf communities, many of people who have these implants find themselves ostracized by both worlds. Some in the deaf community believe that cochlear implants will result in a form of "cultural genocide," although others find the implants to be a relief from a burden.

In comparing the major viewpoints in various communities, it becomes obvious that "normality" relies heavily on perspective. Dorothy Wertz (2002) explored this point by proposing two questions: First, is it ethical to allow hearing parents to abort a deaf fetus? Then, likewise, is it ethical for deaf parents to abort a hearing fetus? Indeed, in a recent survey of deaf parents, several said that they wish to have a deaf child, just as there are dwarf couples who state they would prefer their child also be a dwarf (Dennis 2004). While a large majority of geneticists supported the desire of hearing couples to abort a fetus known to be deaf, they rejected the abortion of a hearing fetus by deaf parents as an exploitation of the power of genetic diagnosis to knowingly bring a disabled child into the world, theoretically at the expense of a "normal" child.

Intersexuality

It is estimated that as many as 1 in every 100 children is born "intersex," with a body that is in some way altered from society's conception of a standard male or female (Intersex Society of North America 2004). The term *intersexuality* encompasses a wide range of characteristics affecting genetics and physiology. The genetic condition androgen insensitivity syndrome, described on page 90, highlights the difficulty in designating some individuals as "normal" or "abnormal."

Most societies have fairly strict views of gender and what constitutes a male or a female, and because people with ambiguous genitalia are in a very small minority and are not capable of reproducing, they are viewed as abnormal. Ambiguous genitalia are often a source of great distress for parents, and confusion about the nature of the condition often leads parents to turn to surgery, hormone therapy, and forced gender roles for their children.

One of the major contentious issues surrounding the "normalization" of children with ambiguous genitalia is the claim that that the surgical and behavioral measures taken are dangerous, unnecessary, and done without the consent of the patient (Dreger 1998; Fausto-Sterling 2000). Of all the intersex conditions, only one—congenital adrenal hyperplasia—constitutes a medical emergency.

Very often a parent's first question is, "Is it a boy or a girl?" This is such a fundamental concern that many physicians feel that it should be answered immediately, and that it is unfair to give parents a sexually "ambiguous" baby. Groups of intersex individuals, however, have disagreed. They claim that before parents even begin to get to know their child, they are making decisions about which gender the child ought to be, and how this gender should be manifest in the behavior of the child. Moreover, because significant sexual brain patterning occurs before birth, these groups say that forcing the child to assume the role of one sex or the other can have severe psychological consequences.

The Intersex Society of North America (2005) advocates that newborns with intersex characteristics should be raised as either boys or girls, depending on which of those genders the child is more likely to "feel" as he or she grows up. But they also maintain that such "gender assignment does *not* involve surgery; it involves assigning a label as boy or girl to a child. … Genital 'normalizing' surgery does not create or cement a gender identity; it just takes tissue away that they patient may want later." There are numerous cases of individuals who were forced into a gender that was at odds with who they were (see Dreger 1998; Colapinto 2000).

The Intersex Society also doesn't advocate the creation of a "third gender" of intersexuality, feeling that to assign a child to an "intersex" gender would probably stigmatize and traumatize the child. Rather, the Society sees male and female not as separate, discrete categories, but as a continuum in which an intersex individual can find their own place. The United Kingdom Intersex Association adds that if the intersex individual does decide as an adult to have "refashioning" surgery, that is the individual's choice, and should not be made for an infant by parents or medical staff.

The notion that intersex individuals should be able to choose for themselves whether to become either male or female is not without controversy. Not all people with intersex conditions are happy with the notion of having to conform to society's definitions and would like to have a separate category. One, Mairi MacDonald, was not satisfied with being male, or being forced into the gender roles associated with maleness. "[G]iven the choice of 'male,' 'female,' 'intersex,' I would unhesitatingly select 'intersex'—but society does not give me that option, so I select 'female.' I do so with deep reservations, gritting my teeth at a society which will not accept my right to simply be who I am" (MacDonald 2000).

The Ethics of Normalcy

The notion of normalcy—who gets to define it and how prenatal diagnosis might enforce it—are going to be ongoing issues in the next decade. As society has become more conscientious about discrimination, experts in the fields of medicine and philosophy have made greater attempts to define normalcy as objectively as possible. Many have lauded functionality as the best guide to normalcy because of its objectivity. Biological function, they argue, is more easily measured and defined than, for example, quality of life (Reznek 1987). Others, however, maintain that measurements of functionality do not remove judgment from the "abnormal" label. After the

objective measurements have been collected and the bell curves constructed, subjective decisions still need to be made regarding where "normal"
stops and "abnormal" begins (Vacha 1985, 1978). And once these decisions
have been made, what impact do they have on those considered abnormal?
Do people with disabilities have lives of lower quality than those considered normal?

Many who do not live with disabilities (or with the disabled), including
members of the general population and many within medicine and academia, would answer this final question in the affirmative. However, surveys
have shown that many people who are considered biologically abnormal
report a high quality of life. Philosopher Dan Brock takes a rather bold position on these findings and claims that personal reports on quality of life are
subjective and therefore invalid (Amundson 2005). His premise is based on
the Boorsian "biostatistical theory" of health. In this view, each body part is
considered normal if it functions at "statistically typical efficiency," defined
as "levels within or above...[a] central region of the population distribution"
for the individual's reference class (Boorse 1977). Based on this analysis,
Brock contends that it is the purpose of health care to maintain normal,
non-impaired functionality—which he asserts entails protecting or restoring normal opportunity as "a necessary condition for a high quality of life"
(Amundson 2005).

Using such an argument, even if a person with a disability claims to have
a high-quality life, the medical world can conclude otherwise based on
physical measurements. The perpetuation of this belief has led to efforts to
"normalize" disabled children with the hopes that their lives would be
enhanced. As reported by Alice Dreger of the Center for Ethics and Humanities at Michigan State University, the efforts often have the opposite effect
(Dreger 1999). In one particular study, girls who were genetically predisposed to short stature were given growth hormone in the hopes that being
taller would help them fit in better with their peers. Instead, the girls—who
originally had not thought much about their height—were made to feel like
freaks because they were singled out for the study and told that they were
abnormally short but that medicine could help make them "normal"
(Dreger 1999). This trend has been seen in numerous other individuals with
disabilities. Their disabilities are simply part of who they are. They do not
feel freakish or abnormal until someone tells them that they are abnormal.
Dreger's position is that this is a problem with society, not with the individuals targeted, and contends that most people with disabilities would rather
like to see society rid itself of prejudices instead of trying to "fix" or eliminate those with disabilities.

Moving along these lines, Eleanor Muirhead and Susan James (1999)
argue that the dichotomy of "normal" and "abnormal" is entirely artificial
and should not be used to describe human beings. They demonstrate how
concepts of normal and abnormal have changed over history by citing

examples from history, such as eras when having red hair could brand a person as a witch. They point out the dangers of assuming that everyone we consider to be abnormal actually wants to conform to the established norm. They posit those with disabilities only feel abnormal when others point it out to them, and that by trying to "fix" the abnormality we are rejecting the individuality of that person.

Many scholars and scientists have rejected the concept of a universal norm. Ronald Amundson has cited the success of "abnormal" individuals as a means of rejecting the notion that abnormal is inherently bad. A striking example features a British student with hydrocephaly, a condition in which the brain is filled with excess cerebrospinal fluid and the volume of brain tissue (gray and white matter) is often greatly reduced. Although this condition often creates serious functional problems, it had virtually no impact on this particular student's life. He had an IQ of 126 and led a normal social life, even though an examination revealed that his brain volume was only about 10 percent of that of most people's. Clearly, his brain was not functioning in the usual way but was able to achieve usual results. Amundson cites this example as evidence that there are multiple ways for natural systems to function successfully (Amundson 2005).

Lennart Nordenfelt has suggested that any characteristic is normal and healthy as long as a person is able to reach their "vital goals," defined as the "states of affairs [that] are necessary for minimal long term happiness" (Nordenfelt 1985). This definition allows flexibility in defining how individuals attain happiness. Different people can have very different expectations and requirements for happiness and therefore there can be no universal standards for what is "normal" or "abnormal."

In conclusion, there are many different ways to define "normalcy," from statistical averages to functionality to level of happiness—but is it necessary to choose one of these definitions? Is it essential that we label some people as "normal" and others as "abnormal?" At any moment, any given social group, regardless of phenotype, could define itself as the "norm" while viewing all other groups as abnormal.

It is interesting, although perhaps not surprising, to note that a large majority of the books and essays on the concept of "normal" are written by people who, by their own admission, are considered abnormal by much of society. Those in the minority are often the first to challenge the conceptions held without question by those in the majority (see Gilbert 1979). Such questioning is among the means that allow human society to move forward. Society is constantly revising what it will accept as normal and what it defines as abnormal. Indeed, that is what is usually referred to as "history."

CHAPTER 14

Genetic Essentialism

DNA in popular culture functions, in many respects, as a secular equivalent of the Christian soul …. In many popular narratives, individual characteristics and the social order both seem to be direct transcriptions of a powerful, magical, and even sacred entity, DNA.

DOROTHY NELKIN AND M. SUSAN LINDEE (1995)

What makes animals look and behave the way they do? What gives each person his or her appearance and personality? In scientific terms, where do we get our phenotype? The answer that we receive both from popular culture and from most science books is unanimous: our genes give us our phenotype. Genes are the basis of heredity, and DNA is the "secret of life," the "master molecule" of genetics. In television shows, in advertisements, in documentaries, and in novels, we are told that DNA makes us what we are physically, mentally, and behaviorally (see Keller 1992; Nelkin and Lindee 1995). A magazine cover shows the canonical double helix and proclaims: "Were you born that way? Personality, temperament, even life choices. New studies show it's mostly in our genes" (*Life* magazine, April 1998). We are told that scientists have found genes for homosexuality, mathematical ability, and criminality (even though they haven't).

The notion that DNA is the core of our existence is so much a part of our culture that we use DNA as a metaphor for that which is essential. Thus a

European newsletter tells us that "the sauna is in the DNA of the Finns." A luxury car is billed as "a genetic superstar," and another claims that "While some luxury sedans just look like their elders, ours has the same DNA." A recent advertisement for the midsize Hummer proclaims "Same DNA. Smaller chromosomes," linking DNA with the car's essence and making a connection between genotype and phenotype (essence and appearance).

One of the arguments used to assert that fertilization is the beginning of an individual human life stems from this centrality of the genome. **Essentialism** is a philosophy that contends that things have core properties that give these things their common, or "essential" identity. The essence of water, for example, is H_2O—water always has two hydrogen atoms that are bonded to one atom of oxygen. Whether the water is liquid or ice, in a pond or in a teacup, does not matter; these qualities are "accidental" as opposed to "essential." Some philosophers have maintained that the essential quality of being human is having a human genome. Because the human genome is formed at fertilization (or, more accurately, during the late single-cell stage), Princeton philosopher Robert George (2002) states the case that

> [T]he hard nugget of the self, a genetic gift … biology is our hidden fate.
>
> CAMILLE PAGLIA (1992)

> Fertilization produces a new and complete, though immature, human organism. … The embryonic, fetal, infant, child, and adolescent stages are stages in the development of a determinate and enduring entity— a human being—who comes into existence as a single cell organism. … The direction of its growth is not extrinsically determined, but is in accord with the genetic information within it.

The notion George expounds here is that of **genetic essentialism**, a doctrine that equates being a human person with having a human genome. **Genetic determinism** is the related view that our genes determine our physical, emotional, and behavioral phenotypes, irrespective of the environment in which we develop and grow.

Although genes are critically important in producing a person's physical and behavioral characteristics, it is not true that the genes alone are responsible for one's phenotype. Certainly, a person will probably resemble his or her biological parents (the source of the genes) and full siblings (who share many of the same genes), and genes play a crucial role in this outcome. But the viewpoint that genes are responsible for all our physical and behavioral attributes depends on an outdated and incorrect belief that the human phenotype is produced only by the genes, and that the environment (George's extrinsic causes) does not play a significant role.

George's philosophical quotation is in many ways a translation of the more extreme positions taken by some current popularizers of biology, such as Richard Dawkins in *The Selfish Gene* (1976) and *The Blind Watch-macker* (1986), who writes of the genome as the book of life and who argues that our bodies are merely transient vehicles for the survival and propagation of their immortal DNA. In an essay called "A Vision of the Grail," Walter Gilbert (1990) likens the genome to each of us owning a compact disc that we can point to and say "Here is a human being; it's me." And there is the rhetoric of James Watson and others of the Human Genome project, who claim that understanding the genome will be to find out who we are and to finally "know ourselves." Following their line of thinking, once we know the sequence of the human genome, we will know how to cure such diseases as homelessness (Koshland 1989) and homosexuality (Watson 1997).

But any parent of "identical" twins knows that genetic essentialism is a gross oversimplification. Identical twins have exactly the same genes, but they can be, and usually are, very different people. Even Eng and Chang Bunker, the most famous of the conjoined ("Siamese") twins—who shared not only the same heredity but also the same environment—grew to become different people. One was morose and enjoyed his liquor, while the other was cheerful and abstained from drink (see Gould 1997).

Scientific evidence supports what twins' parents know. Ecologists, embryologists, and natural historians have long known that the genome an organism inherits is responsive to the environment both during and after gestation. What is inherited is a genome whose phenotypes depend greatly on environmental circumstances. In ecology and developmental biology, this ability to respond to the environment by altering one's development is called **developmental plasticity**. Indeed, this interaction between genes and environment to produce the phenotype has been a major part of classical genetics, from Wilhelm Johannsen's first use of the word "phenotype" in 1903, to Richard Lewontin's analysis of the phenotype 100 years later (see Lewontin 2001).

Developmental Plasticity: Examples from the Animal Kingdom

Environmental sex determination

Although sex determination in mammals is genetic (see Chapter 5), there are many animal species in which sex is determined by the animal's environment. Embryos of the turtle *Emys*, for instance, become males if the eggs are incubated below 25°C, and females if incubation is above 30°C during the last third of their incubation (Pieau et al. 1994). In intermediate temperatures, different percentages of both sexes are formed. The specification of sex here appears to be regulated by enzymes that metabolize sex hormones differently at different temperatures.

The sex of the blueheaded wrasse (*Thalassoma bifasciatum*) is determined by the social structure into which the immature fish enter. If a young fish enters a coral reef where there are no males, it becomes male, but if a male wrasse is present, it becomes a female. Usually, there is one male for about a dozen females. When that male dies, the largest female develops testes within 24 hours, and within two weeks it is making functional sperm and mating with the remaining females (Warner 1993). For another coral reef fish, the clownfish or anemone fish (*Amphiprion* species, immortalized in the Disney movie *Finding Nemo*), the situation is reversed. The largest clownfish living in an anemone will be the only female; the other dozen or so fish in the group are males. If the female dies, the largest male clownfish changes to become the group's only female.

The sex of some invertebrates, such as the slipper snail *Crepidula fornicata*, is context-dependent—it depends on what the animal settles down on. Slipper snails are all born male. The young snails pile up on top of one another, and an individual snail's sex is determined by the sex of the snail it is attached to (it will become the opposite). Another invertebrate, the pillbug *Armadillium vulgare*, can have its sex determined by bacterial infection. If certain bacteria are present in the egg cytoplasm, they can turn genetic male pillbugs into females. Ecologists have been aware of these environmentally driven life history strategies for years, and they have proposed some interesting hypotheses for why they evolved (see Gilbert 2001, 2003).

Physical characteristics and the environment

Sex isn't the only phenotypic trait influenced by the environment. The eighteenth-century naturalist Carolus Linnaeus, who developed the scientific classification of organisms using Latinized genus and species names, classified the brown and the gold butterflies shown in Figure 14.1 as two different species. We now know that these two phenotypes are really two forms of the same species (the European map butterfly *Araschnia levana*), and that the wing coloration is regulated by the temperatures and amount of sunlight experienced by their larvae (i.e., caterpillars). More daylight and higher temperatures cause higher amounts of ecdysone (an important insect hormone) and produce the brown summer morph. Less daylight and lower temperatures produce the orange-gold spring morph (Nijhout 1991). Such seasonal shifts are seen in numerous butterflies, including the common cabbage white (*Pontia* species), in which butterflies emerging in the spring have dark hindwings that absorb more light and raise the body temperature faster (giving them an advantage in cooler temperatures).

Several vertebrates and invertebrates change their development when a predator is sensed (see Gilbert 2001). By sensing a predator-secreted chemical in the environment, some species of crustaceans, molluscs, fish, and reptiles develop differently. The carp *Carassius carassius* responds to the

FIGURE 14.1   European map butterflies that emerge from their cocoons during the summer are predominantly brown (top). Those that emerge in the spring have an orange-gold color pattern (bottom). Early naturalists believed these "morphs" to be two different species, but the color difference is actually a result of hormone levels, which differ based on the temperature and amount of sunlight (photoperiod) the larvae are exposed to. (Photographs courtesy of H. F. Nijhout.)

presence of predatory pike by developing a pot belly and a hunchback, which do not fit well in the pike's jaws The predator-induced form of the water flea *Daphnia cucullata* is beneficial not only to itself, but also to its offspring. When *Daphnia* encounter chemicals produced by larvae (maggots) of the fly *Chaeoborus*, their spiked "helmets" grow to twice their normal size (Figure 14.2), which helps protect them from being eaten by the maggots. The offspring of a water flea with an induced helmet will be born with this same altered head morphology (Agrawal et al. 1999).

In some cases, you are what you eat. In numerous species of wasps, ants, and bees, the worker, soldier, and queen castes are determined by the amount of food fed to the respective larvae. The differences among members of each caste are both physical and behavioral. Similarly, the appearance and behavior of the male dung beetle depends on the dung a larval beetle is fed by its mother (Emlen 1997). Male larvae fed sufficient high-quality dung develop horns, the size of which is in proportion to the quantity of dung eaten (Figure 14.3A); as adults, these

FIGURE 14.2   *Daphnia*, also known as water fleas. These micrographs show an individual who developed in the presence of biochemicals released by a predator (left) and another who developed without exposure to such chemicals (right). The pointed "helmet" is a protective trait that makes it more difficult for the predator (a maggot, or larval fly) to swallow *Daphnia*. (Photographs courtesy of A. A. Agrawal.)

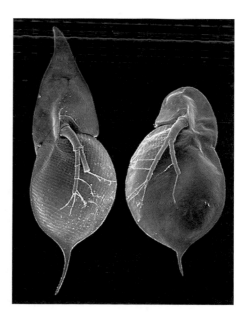

FIGURE 14.3   Two male dung beetles at the same developmental stage. The size of the horns depends on the larval diet. Larvae fed large amounts of dung develop large horns (left); those whose dung diet is less nutrient-rich do not develop horns, and as adults they have different mating behaviors than horned males. (The blue coloration was artificially added to these micrographs to emphasize the horn structure.) (Photographs courtesy of D. Emlen.)

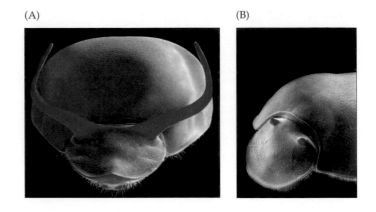

(A)   (B)

males fight for mates. Those males who as larvae did not receive such high-quality dung lack horns (Figure 14.3B), and obtain mates by digging tunnels that "sneak past" the horned males.

These are only some of the many other examples of developmental plasticity that enable developing organisms to respond to changing environmental conditions (see also Adler and Harvell 1990; Tollrian and Harvell 1999). None of these phenotypes is the direct result of the individual's genes. Rather, the genes give the organism the ability to respond to environmental cues.

The Insufficiency of the Genome to Complete Development

It is well established that the environment plays important roles in forming the phenotypes of many animals. Does this have any relevance to humans? Certainly the human species has enormous developmental plasticity. Furthermore, there is evidence from both humans and other mammals that the genome alone is not sufficient to allow us to complete normal development.

Connective tissue malformation

One interesting case of environment effecting our phenotype involves the disease gulonolactone oxidase deficiency, also known as hypoascorbemia, or scurvy. A homozygous mutation (i.e., a mutation in both copies of the gene) of the gulonolactone oxidase gene on chromosome 8 produces a syndrome that involves connective tissue malfunction and results in early death. Amazingly, this syndrome effects *100 percent* of the human species—that is to say, everyone.

Gulonolactone oxidase is the final enzyme in the pathway leading to the synthesis of ascorbic acid, better known as vitamin C. Most mammals have functional genes for this enzyme and thus can synthesize vitamin C; however, all humans are homozygous mutants and cannot make this essential compound. Humans must obtain vitamin C from the environment—from the mother's diet in the case of the fetus, or from food or supplements in the case of the mature organism. Without this "replacement therapy" from the environment, we would all be dead. This situation shows that the effects of genes and environment are interactive; they cannot be separated into component percentages. Our genotype programs us for an early death, and the fact that we are here at all is testimony to the power of the environment to circumvent our genetic heritage.

Intestinal development

Another example of the insufficiency of the human genome to complete normal development concerns our intestinal bacteria. Recent evidence shows that the gut bacteria actually regulate mammalian intestinal genes, and these symbiotic bacteria are necessary in order to complete intestinal development. At least this is the case in mice, and it is probably true in humans as well. Researchers found that a particular enzyme characteristic of mouse intestine is induced by bacteria (Umesaki 1984), and further studies have shown that the intestines of bacteria-free mice can initiate, but not complete, their differentiation (Hooper et al. 1998).

Analysis of mouse intestinal cells has shown that naturally occurring bacteria can increase the transcription of several mouse genes, including those encoding the proteins colipase, which is important in nutrient absorption; angiogenin-3, which helps form blood vessels; and sprr2a, a small protein that is believed to fortify supporting tissues that line the intestine. Without a particular intestinal microbe, *Bacteroides thetaiotaomicron*, the tiny capillaries of the small intestinal villi fail to develop completely (Stappenbeck et al. 2002).

Another pair of bacteria—*Bacteroides fragilis* and *Bacillus subtilis*—appear to be needed together in order to induce the formation and maturation of the gut-associated lymphoid tissue, or GALT (Perry and Good 1968; Cebra 1999). The lymphocytes of GALT mediate immune protection in the mucous lining of the intestines and allow us to ingest food without mounting an immune response (allergic reaction) against it. In humans, GALT malfunction has been associated with allergies and with Crohn's disease, a serious inflammatory bowel syndrome (see Rook and Stanford 1998; Steidler 2001). When introduced into germ-free rabbit appendices, neither *B. fragilis* nor *B. subtilis* alone was capable of inducing GALT; however, the combination of these bacteria consistently induced the formation of these gut lymphocytes (Rhee et al. 2004).

Interestingly, bacterial symbiosis in gut tissue formation is not limited to mammals. Analysis comparing gene expression in the digestive tracts of germ-free and conventionally raised zebrafish also show that bacteria are needed to induce some gene expression. In fact, there are at least 200 microbially induced genes in the gut of the fish, including several that are involved in epithelial cell proliferation, nutrient metabolism, and immune response (Rawls et al. 2004).

The genome is truly remarkable. It serves to connect each generation through a set of organ-forming and species-specific instructions. But one of the most incredible properties of the genome is that it lets individuals respond to their environments by changing their development.

The Insufficiency of the Genome to Determine the Phenotype

If we know the genotype, can we accurately predict what the phenotype will be? The answer to this question is "no" (see Gilbert 2001). One reason the genotype does not predict the phenotype is that differential DNA methylation modifies gene expression, and this methylation can be affected by the environment. In addition, both random chance and the capacity of humans to learn can have significant effects on the phenotype.

Methylation

Recall that in Chapter 7 we described how DNA is modified by methyl groups, and that when these methyl groups are present on a DNA sequence, they prevent the gene from being active. Some of this methylation is the result of nuclear factors present in the sperm and the egg. However, methylation is also directed by the environment. Moreover, this environmentally induced methylation can alter the phenotype in very substantial ways.

PIGMENTATION   Experiments by Waterland and Jirtle (2003a) demonstrated that differences in the diet of pregnant mice can produce differences in the DNA methylation of their fetuses, and that these methylation changes can affect the pup's phenotype—in this case, coat color. They fed pregnant yellow mice dietary supplements, including folate* and choline, that provided high amounts of methyl groups. The researchers found that the more of these supplements the mother received, the greater the methylation of a specific gene in their offspring. The gene involved produces an enzyme that turns dark pigment yellow; methylation turned the gene off, and the pups' pigment remained dark brown. Thus, the more folate the pregnant mice were fed, the darker their offspring's fur coats.

---

*Folate, or folic acid, is one of the B vitamins. Obstetricians usually recommend a folate supplement for pregnant women, as it has been shown to help protect the human fetus against neural tube closure defects (see Figure 1.16).

BEHAVIOR   Another study found an even more remarkable effect of the environment on phenotype. It has been known that rats who are licked and groomed by their mothers during the first week of birth tolerate stress much better as adults. An excess of stress hormones—the "fight-or-flight" chemicals, such as glucocorticoids (cortisol) and adrenaline, released in response to dangerous or stressful situations—in the mammalian bloodstream is detrimental to the individual's overall health. Stress tolerance in rats has been linked to the presence of a certain protein receptor (the glucocorticoid receptor) in the brain of adult rats. If this receptor isn't present, the rats are less able to rid their systems of excess stress hormones (Weaver et al. 2004). Experiments revealed that the difference between the rats who were licked and groomed and those who weren't so well cared for was the methylation of a particular region of the glucocorticoid receptor gene. In rats that received maternal care, the DNA was unmethylated, so the pup formed glucocorticoid receptors in its brain and thus responded well to stress. If the rat was neglected, the DNA became methylated, the glucocorticoid receptor levels in the brain were low, and the rat later responded poorly to stressful situations.

DISEASE SUSCEPTIBILITY   Epidemiological and anatomical data from humans (reviewed in Barker 1994a,b; Jablonka and Lamb 2002; Gluckman and Hanson 2005) have indicated that prenatal nutrition influences the adult offspring's susceptibility to cardiovascular disease, obesity, cancers, and type 2 diabetes. Waterland and Jirtle (2004) provide evidence that DNA methylation might be the critical link by which nutritional deficiencies during a woman's early pregnancy might cause chronic disease in her adult offspring. Thus, environmental circumstances during our development may play critical roles in determining our physical and behavioral phenotypes, as well as possibly making us more prone to have certain diseases as adults.

Chance variation: Lessons from twins

Much has been made about the Minnesota Twin Studies, which studied pairs of adult identical twins raised apart and other pairs raised together. Popular accounts of these studies tell readers that some identical twins raised apart since birth share remarkable common behaviors, such as flushing the toilet twice, wearing large belt buckles, and marrying spouses with the same first name. However, neither a "diploflush" nor a "macrobuckle" gene has ever been isolated, and a survey of whether any large fraction of twins marries spouses of the same name might show no correlation at all. Certainly, there doesn't seem to be any correlation in this regard among twins raised together.

Identical twins—with the same DNA in their genome—are very likely to be *discordant* in behavioral phenotypes (see Taylor 2004; Lindee 2005).

Indeed, identical twins reared together often distinguish themselves by purposefully going in different directions and having different behaviors, a response that can be viewed as a result of the presence of an identical sibling in an individual's environment (Farber 1981).

In females, one of the two X chromosomes becomes inactive in each somatic cell (see page 89); this equalizes the number of functional X chromosomes in male and female cells. However, there is no determining *which* X chromosome will be inactivated in any given female cell. This randomization of X-chromosome inactivation can give rise to identical (monozygotic) female twins who are profoundly different. Take, for example, the case of identical twins who are heterozygous for an X-linked form of muscular dystrophy (see Chapter 5). Most heterozygous women do not express muscular dystrophy symptoms because the cells expressing the normal allele can compensate for those cells expressing the mutant allele. However, if by chance the normal allele is on the inactivated X chromosome in a large proportion of a woman's muscle cells, that woman will manifest the disease. There have been several instances where one twin shows the symptoms of this disease while her identical sister does not (Norman and Harper 1989; Richards et al. 1990). There are similar cases where monozygotic female twins are discordant with respect to color blindness as a result of X-chromosome inactivation. Thus, while genes play an extremely important role in creating the phenotype of each human, the phenotype cannot be predicted by the genotype.

Learning

We are constantly exposed to supposed examples of genetic determinism. We are told that Ringo Starr's child has "drummer genes," and that figure skating champion Yuka Sato gave her daughter a "genetic advantage" toward becoming a champion herself. However, for all the newspaper stories, we have never identified genes for drumming or figure skating ability. Nor have we found genes (if there are any) that inevitably produce homosexuality, shyness, aggressive behavior, schizophrenia, manic-depressive psychosis, rhythm, altruism, or television viewing (see Mann 1994; Harris 1997).

What we *do* find is actually much more interesting. The plasticity we discussed above in relation to the physical phenotype is even more pronounced when we look at the human nervous system, where plasticity creates human individuality. It appears that new neurons are being generated by experience, and that neural connections are constantly being formed, retained, or destroyed as a result of experience.

If there is any trait that distinguishes human development from that of the rest of the animal kingdom, it is our brain structure. Specifically, humans are primates who retain the immense neural growth rate of a fetus even after they are born. During the first months after birth, a human being adds approximately 250,000 neurons *per minute* to their central nervous systems (Purves and Lichtman 1985). Apes' brains have a similarly high growth rate before birth, but this rate slows greatly after an infant ape is born.

Humans experience rapid brain growth both before birth and for about 2 years thereafter (Martin 1990). At birth the ratio of brain weight to body weight is similar for great apes and humans. By adulthood, however, the human brain-to-body-weight ratio is 3.5 times larger than that of apes (Bogin 1997). It is thought that the immense increase in the number of neurons and neuronal connections in the human brain may (1) generate new modules that can acquire new functions, (2) store new memories for use in thinking and forecasting possible scenarios, and (3) learn by interconnecting among themselves and with those neurons generated before birth (Rose 1997). Moreover, some studies document that interventional and enrichment during these first 2 years can raise a child's intelligence quotient (reviewed in Wickelgren 1999).

The prodigious rate of continued neuron production in young humans has important consequences. Several authors have made the claim that we are essentially "extrauterine fetuses" for the first year of our life (see Portmann 1941; Gould 1977; Montague 1989). These authors argue that we become human through our early interactions with a rich environment of sounds, touches, smells, and sights. Indeed, if you follow the charts of ape maturity, the human gestation period would be 21 months instead of the current 9—developmentally, humans are at 21 months what other apes are at birth. Our premature birth is a compromise between the mother's pelvic width and the baby's head circumference and lung maturity (see Chapter 1).

One of the more interesting speculations concerning the consequences of this retained rapid growth rate is that this extrauterine fetal growth rate necessitated the invention of childhood. Anthropologist Barry Bogin has suggested that *childhood* is a novel stage of the vertebrate life cycle, unique to humans and interposed between the infant and juvenile stages. Infancy ends with weaning, usually around the age of 3 for humans. The juvenile stage is characterized by feeding independence from adults and the onset of physical maturity. In humans, this usually begins around year 7. Between the ages of 3 and 7 in humans, there is a period of childhood, characterized by immature dentition (tooth development), a small digestive system, and a rapidly growing brain that demands a high-calorie diet (Bogin 1997). Childhood is a period of time when the human has to be cared and fed by adults.

The childhood period allows the growing brain to develop in an enriched environment; but Bogin argues that this is actually a byproduct of childhood and not its intention. The selective value of childhood would be to improve the success rates of each child surviving to maturity. This would explain why humans have lengthy development and low fertility—both of which would seem to be detrimental to evolutionary success—and yet have the greatest reproductive success of any ape species. As human geneticist Barton Childs (1999) has succinctly written, the retention of fetal characteristics in the newborn "allows us to escape the tyranny of our genes, or at least temper their rule."

There is extensive evidence that experience creates more neurons and more neuronal connections; the brains of young rats reared in stimulating environments have been shown to be packed with more neurons and connecting neural tissue than are found in rats reared in isolation (Turner and Greenough 1983). Even the adult brain develops in response to new experiences. When adult rats learn to keep their balance on narrow dowels, certain neurons in their cerebellums develop new linkages, or synapses (Black et al. 1990). Studies on rodents have also shown that the neural stem cells from the brain's hippocampus can be cultured in vitro and then grafted back into different regions of the adult rodent brain, where they can respond to specific cues and differentiate into different, site-specific types of neurons (Gage et al. 1995; Suhonen et al. 1996).

> The interaction of individual animals and their world continues to shape the nervous system throughout life in ways that could never have been programmed. Modification of the nervous system by experience is thus the last and most subtle developmental strategy.
>
> D. PURVES AND J. LICHTMAN (1985)

It is thus obvious that there are significant functions of our brains that are not "hard-wired" by our genes. Even in such important areas as vision and sensation, neural connections can be modified, and even the adult brain is still plastic. It is probable that critical adult functions such as learning and memory arise from the establishment and strengthening of different connections within our central nervous systems as a result of new experiences.

Developmental biology has played a major role in critiquing those biological hypotheses that propose the gene or genome to be the ultimate unit of life. Human life is far richer than mere genetic endowment and is all the more remarkable for allowing us to change through experience.

Back to the Essentialist Argument

We should be careful when we hear the argument that all of an individual's developmental program exists once the genome comes into being at fertilization. In fact, there are many entities that have a uniquely human set of genes in them but which aren't considered potentially human (see Frost 2004). The fact that each of our skin cells carries a unique set of human genes does not make skin cells potential humans. We certainly can't claim them as deductions on our income tax. For that matter, neither skin cells nor embryos are considered as people for purposes of taxation or state representation. Having a nucleus with a human genome may not be the "essence" of a person, but may be more of a *sine qua non*—a necessary but not a sufficient cause. All persons must have such a nucleus, but simply having it does not make that entity a person.

The equation of human essence to DNA reduces humanity to its mere biological properties and stands opposed to the more traditional views that human nature is characterized by loving, caring, and other actions that are the product of both hereditary endowment and cultural experience. Sociologists Dorothy Nelkin and Susan Lindee (1995) have pointed out that the essentialist reduction of human essence to DNA so touted by today's popular media is very similar to the Medieval depiction of the human soul. It is seen as the essence of a person, the "true self" that exists independent of the physical body; it determines that person's character and behavior; and it is that immortal substance from which you can be resurrected after death (as described for dinosaurs in *Jurassic Park*). The equation of the human essence and the human genome may be just such a vision of ensoulment. It certainly is not science.

CHAPTER 15

The Ethics of Animal Use in Research

> The day may come, when the rest of animal creation acquire those rights which never could have been withheld from them but by the hand of tyranny … The question is not Can they *reason*? Nor Can they *talk*, but Can they *suffer*?
>
> JEREMY BENTHAM (1789)

All of the scientific fields and medical technologies discussed in this book have one feature in common: research on animals has played a prominent role in their development and advance. Worldwide, an estimated 41 to 100 million animals are used each year in scientific experiments. A handful of countries have laws and regulations protecting animal welfare, but many people argue that the laws don't go far enough. Animal rights and animal liberation movements advocate the abolition of all animal experimentation, asserting that the use of non-human animals as means to human ends cannot be morally justified. Advocates of animal experimentation argue that without the use of animals in research, progress in medicine would halt and the safety of new products could not be sufficiently tested. The debate is a bitter one, but advances in technology and our increased understanding of animal lives have led to at least partial solutions respecting the concerns of both sides.

A Brief History of Animals in Research

Homo sapiens have always depended on other animals for food, clothing, labor, and companionship. The death of animals in the cause of human survival is a fact of existence, just as is the death of prey eaten by any other animal. However, philosophers have debated the ethics of using animals for research and experimentation for many centuries.

In the fourth century BCE, Aristotle justified human use of animals by asserting that animals did not possess a rational soul, and that "Nature has made all animals for the sake of Man." One of Aristotle's successors, Theophrastus, raised an opposing voice, claiming that animals possessed at least limited reasoning skills (DeGrazia 2002). In the second century CE, as religious conviction rose against the use of cadavers in medical science, the Graeco-Roman physician Galen began the first animal experiments in the name of medicine (Regan 2001; Phillips and Sechzer 1989). The Middle Ages brought reaffirmation of Aristotle's views by philosophers and theologians alike, and most Christians have since upheld the assertions of Thomas Aquinas that animals cannot reason, and are therefore subordinate to humans (DeGrazia 2002).

Among Western philosophers, René Descartes (1596–1650) is the most vilified by today's animal rights movement. Descartes was perhaps the most important proponent of the emerging Renaissance ideal of knowledge based on reason and evidence rather than religious faith and blind acceptance of the classical masters such as Aristotle. Although Descartes believed that animals had feelings, which they expressed through sound, he did not believe that they suffered as humans did. He advocated vivisection—the unanesthetized dissection of living animals—in the interest of obtaining evidence and advancing knowledge. Although, very few scientists fully advocated Descartes' views on this point, and the Catholic Church condemned his writings a decade after his death, vivisection did take place in Renaissance Europe.

The use of animal models in research came into the Western mainstream early in the nineteenth century. Physiologists Claude Bernard and François Magendie guided many European scientists toward the belief that medical research could be conducted only through animal models (Phillips and Seczher 1989; Cohen and Regan 2001; DeGrazia 2002). Also around this time, the first animal rights activists began to voice their concern about the humane treatment of animals, a cause that grew in prominence as a moral issue over the course of the nineteenth century. Animals took on human qualities in books such as Anna Sewall's famous *Black Beauty* (1877), in which the hero and narrator is a horse. Pets became more and more common as members of urban households, and the idea of Darwinian evolution blurred the line between humans and other animals (Sperling 1988). Other moral concerns were also coming to the forefront, and antivivisectionist literature often made references to child protection, temperance, feminism, and pacifism. The vast majority of antivivisectionists of the era lived in cities, and most belonged to the upper-middle class of professionals and clergy. Although these urban-based groups targeted many of their publications to the working and rural masses, large antivivisection movements in these classes were never sustained (French 1975).

Women played a prominent role in the Victorian antivivisection movement, connecting themselves to the animal cause through concepts such as feminine nurturing instincts and emotionality; their opponents often ascribed negative and stereotypically feminine traits such as irrationality and hysteria to the antivivisection movement (Sperling 1988). Antivivisectionist leaders such as Anna Kingsford and Frances Power Cobbe made attempts to indirectly further the feminist movement by urging women to become politically involved with antivivisection (French 1975). Animal rights movements today are still strongly linked to feminism, as well as nature-oriented and humanist movements such as holistic healing, the antinuclear movement, and environmentalism (Sperling 1988).

With the passage of the Cruelty to Animals Act in 1876, Great Britain became the first country to enact national legislation to protect animals. (The United States did not enact animal protection laws until 1966—nearly a century after England's original legislation.) The Act diminished the number of animal experiments performed, but did not weaken the opposition to the practice. In the remaining years of the nineteenth century, animal protection groups formed and grew, demonstrations erupted, and new animal protection bills flooded Parliament. This surge in activity led to some new legislation and increased awareness among scientists, the government, and the public (Phillips and Sechzer 1989; Orlans 2002).

By the beginning of the twentieth century, the antivivisectionist movement had lost momentum and all but disappeared. Then, as Hitler gained power in Europe during the 1930s, antivivisection found a host of unlikely supporters. With the rise of the Nazis in Germany came a new set of incredibly detailed laws and regulations that placed tight regulations over everything from hunting, to animal experimentation, and even the method of killing lobsters in a restaurant (Arluke and Sanders 1996). Called "a tragic irony of the satanic values of the Nazi regime" (William Seidelman, quoted in Phillips and Sechzer 1989), this legislation was established to protect domestic animals, research animals, and wildlife from harsh treatment and pain. This paradoxical Nazi compassion toward animals is attributed mainly to anti-Semitic misrepresentations of Kosher slaughter practices as well as to propaganda techniques. Hitler himself gained support by touting his vegetarianism and compassion for animals (Arluke and Sanders 1996).

The vegetarianism of Hitler and many other Nazi leaders was also a product of racism rather than compassion. They believed that the blond-haired, blue-eyed Aryan race was superior to all others, but had been contaminated by intermixing with other races and by eating meat. Their solution: eliminate meat from the diet and non-Aryan people from the planet. Some of the impetus behind the Nazi animal protection laws might also be explained by a cult-like worship of dehumanized nature. National Socialist ideology decried human compassion as insincere, and touted instead the obedience of pets and the fearless cruelty of the predator as behaviors to emulate (Arluke and Sanders 1996).

The German animal protection legislation was erased after World War II, along with all other laws enacted by the Third Reich (Phillips and Sechzer 1989). In fact, with the world horrified by Nazi use of humans in their medical studies, animal use skyrocketed after World War II. The U.S. Food and Drug Administration began to require animal testing for all drugs developed for human consumption. These new regulations, coupled with generous government funding for biomedical research, led to an enormous increase in the number of animals used in scientific studies (DeGrazia 2002). In the mid 1950s, more than 150,000 rhesus monkeys were imported into the United States to be used for research and testing. Some 120,000 of these monkeys were killed in the development of the Salk polio vaccine (Haraway 1989).

Beginning in the 1960s, there was a shift in animal philosophy away from behaviorism—the idea that behavior is completely separate from psychology—toward the new field of cognitive ethology. Pioneered and publicized by Donald Griffin, cognitive ethologists found that animals experience pleasure and pain, and that many animals can act out of intention rather than completely instinctively, as previously thought (DeGrazia 2002; Bekoff 2002a). Awareness of animal welfare began to rise again, and in 1966 the United States passed the Animal Welfare Act (AWA). The act mandated larger cages, required anesthesia for painful procedures, and established Institutional Animals Care and Use Committees (IACUCs; see page 255) so that institutions could monitor their own animal use (U.S. Senate Committee 1966).

The Animal Welfare Act does not cover farm animals, reptiles, amphibians, fish, birds, mice, or rats. These last two species make up the vast majority of animals used by biologists and physicians, and government officials cite the rising costs of inspections as justification for their exclusion from protection (Regan 2001). Little has been added to the act since its inception, and in 2002, the latest reexamination of the AWA, Congress voted to block its expansion to include birds, mice, and rats (Gewin 2002).

Concern for animal welfare increased still further during the 1970s. Michael Fox attributes this to increased knowledge of animal psyches, deeper awareness of the interconnectivity of ecosystems, and expanding moral concern stemming from the civil rights movement and anti-war activism of the 1960s (Fox 1986). Peter Singer's publication *Animal Libera-*

tion (1975) fueled the animal welfare fire, and in 1980 Henry Spira exposed the public to the horrors of two toxicity tests used by the U.S. chemical industry. The first of these tests, the LD50, or lethal-dose 50, test force-feeds a set of animals a product until 50 percent of them die. The second test, the Draize test, involves applying a product to rabbits' eyes until massive tissue damage is achieved. Because of Spira's work, the use of both tests has been scaled down considerably, and as of 1983, the FDA no longer required LD50 data (DeGrazia 2002).

During the 1980s, the United Nations World Health Organization (WHO) instituted a basic international framework, the International Guiding Principles for Biomedical Research Involving Animals (CIOMS 1985), that countries can accept voluntarily. At least 23 countries have instated their own animal welfare legislation. Basic animal husbandry, inspection of animal holding facilities, and control of animal suffering are the most rudimentary regulations and tend to be the first instituted by governments. More advanced laws cover review of proposed protocols, specifications for investigator competency, bans on certain procedures, animal sources, or species used, and application of the "three Rs" (see pages 257–260). Several countries have placed bans on animal sources, species, and procedures. For example, Germany has banned the use of animals in testing weapons, cosmetics, tobacco, soaps, and alcohol. The Netherlands requires lab animals to be purpose-bred (meaning they must be bred in the laboratory and cannot be taken from shelters or the wild), and the United Kingdom has banned the use of chimpanzees in research. There are no bans on species, sources, or procedures in the United States (Orlans 2002).

Do Animals Have Moral Standing?

Do non-human animals have moral standing, and if so, how does it relate to the moral standing of human beings? Although some people will argue that non-human animals have no moral standing whatsoever, most believe that animals should be granted at least some consideration, though reasons for this argument vary widely. Surely it is less moral to tear apart a laboratory mouse than to tear up a piece of paper. Is it enough that the mouse is alive? Bacteria are alive, but activists do not vilify scientists for working with bacteria, and (if they have any sense) they don't try to liberate bacteria from laboratory cultures.

Why do we teach our children not to burn ants under a magnifying glass or kick the dog or pull the cat's tail? Are we afraid that a child who torments animals has evil tendencies that, if not curbed, will grow and eventually be directed against people? Do we feel an obligation to the cat as a member of the family? Or do we feel that the cat itself has an interest in not being hurt, and it is for the animal's sake alone that we stop the tail pulling? Probably there is a combination of all these factors in our reasoning.

A key issue in the debate on human responsibility to animals is the language used to argue various points. Where two people may agree that animals should not be confined to small cages, one may assert that animals have the *right* not to be confined in tiny cages, while another may say that animals have no rights whatsoever, but that humans have an *obligation* to treat animals well and not to confine them in tiny cages. Many pages of text have been devoted to the various legal and technical definitions of rights that may or may not apply only to humans, and whether a human's "duty" or "moral obligation" to an animal implies that that animal has "rights" or "moral standing" (see Rodd 1990; Singer 1975; Francione 2000; Wolfe 2003; Hearne 1991; DeGrazia 2002).

Utilitarianism

The English philosopher Jeremy Bentham (1748–1832) established the doctrine of utilitarianism, whereby an action is considered moral if it increases the happiness, or "good," of the greatest number of people. Bentham believed that non-human animals also strive for happiness, and that their happiness also counts in the "equation" of utilitarianism. This principle has been reasserted by Peter Singer (1975, 2004). His book *Animal Liberation* was very influential, calling for the release of animals from the "bondage" of the meat and biomedical industries.

Singer argues that no single characteristic separates all individual humans from all other animals, nor do we need to find such a characteristic. His philosophy—which he calls "preference utilitarianism"—asserts that all creatures have interests, and that all creatures' interests count equally (Singer 1975, 2004). Furthermore, for any outcome to be morally justified, it must provide the greatest overall benefit. It follows that a human's interest in eating meat cannot override the interest of a cow to move about freely and to live peacefully until it dies naturally. R. G. Frey, also a utilitarian, points out that it is hypocritical to believe that the ends do not justify the means when it comes to using human subjects for medical research, but it is all right to do experiments that involve animals. If it is not morally right to perform various experimental procedures on humans, he believes, their use in animals cannot be justified either (Frey 2002).

Other writers find flaws in the utilitarian approach. Animal rights activist Tom Regan asserts that all interests are certainly not equal, and offers the example of a gang rape. A utilitarian, he argues, would count the interests of the rapists as equal to the victim's interest of not being raped, and the more people there were involved in perpetrating the rape, the more justifiable it would become in the utilitarian view. Moreover, Regan contends, the utilitarian approach should support the killing of any person who places a burden on society—the elderly or the terminally ill. This cannot be morally correct, he says, and if a utilitarian defends against this charge by saying that an interest in life trumps all other interests, he or she is only admitting a fundamental flaw in the philosophy (Regan 2002).

Regan and others also note the difficulty in measuring the total utility of any act, and argue that widely varying conclusions can be supported depending on how "utility" is measured (Fox 1986; Regan 2002).

Moral agency

Another approach some have taken is to point out that humans are members of a moral community. That is, they understand the concept of morals and have moral obligations to one another. Because animals do not belong to this community, and feel no moral obligation toward humans, humans have no reciprocal moral obligations to animals (see DeGrazia 2002; Cohen 2001; Robb 2002; Fox 1986). Opponents of this view concede that animals may not have moral obligation to humans, but that this is no reason to deny them moral consideration (Regan 2002; Frey 2002). Regan points out that not all humans have a concept of morality, including the severely mentally disabled and infants. Indeed, it has been shown, that human brain development does not allow for full moral understanding until adulthood (Bower 2004). Juvenile courts do not respond to this finding by discounting the interests or rights of young offenders, but instead they extend further protections to the children who cannot yet entirely comprehend the moral world.

Arguments against the moral agency position have been furthered by recent studies showing an understanding of fairness among Capuchin monkeys, who get upset when a neighboring monkey receives a better reward for performing the same task (Milius 2003). Moreover, juvenile coyotes that "cheat" in playing with their peers have a hard time finding playmates later, and a hungry rhesus monkey will not take food if this action results in another monkey being given an electrical shock (Bekoff 2002b). If fairness is considered as one of the underlying tenets of morality, it seems that some animals can be invited at least into the foyer of the human moral house. Some philosophers continue to insist, however, that this evidence of the evolution of morality does not make any difference (see Cohen 2002).

Social contract theory

Social contract theory is similar to the moral agency approach in that those with the moral authority make the rules. In this philosophy, also known as contractarian theory, moral rights are protected by a contract, but only the interests of the contractors are protected. Non-contractors' interests can only be protected if they are in direct line with the interests of the contractors (Regan 2002). For example, you cannot shoot a contractor's cow, because that would damage the contractor's property or feelings, or both.

The interests of the cow in not being shot have no bearing in the matter, and there are no rules about what groups, human or otherwise, are to be included in the contract. Detractors of this form of contract theory insist that it can condone the most blatant forms of discrimination.

Indirect duty

The moral obligations to another's property outlined in social contract theory represent another school of thought espoused by Imanuel Kant. Kant agreed with the philosophical view that although animals lack rights, they cannot rightly be destroyed, injured, or stolen as that would indirectly infringe on the rights of the humans who owned the animals. In addition, Kant warned that cruelty directed at animals would doubtless lead to cruelty to humans (Fox 1986; DeGrazia 2002), and indeed, there are data to support this assumption (Muscari 2004; Wright and Hensley 2003; Gleyzer et al. 2002). Other proponents of indirect duty discourage harm to animals because we depend on them for our survival (Robb 2002); abused animals will be less helpful to humans in terms of providing either labor or scientific data, and the abusive act may evoke negative feelings in those people who see or hear about it (Fox 1986).

Social bonds

A theory developed by Mary Midgley (1983) may explain in part why humans are so reluctant to extend rights to other animals. This social bonds theory focuses on relationships, rather than social status, as the driving force behind moral obligation. Strong relationships (i.e., to your family and local community) beg greater moral obligation than weak relationships (i.e. strangers, citizens of other countries, and members of other species). Donna Haraway also believes that moral obligations to animals arise from our relationships with them, but takes the position that, although we naturally prioritize the moral standing of our own species above others, our obligations to other animals cannot be casually discarded (Haraway 2003). As David DeGrazia (2002) analyzes the theory, it can be used to defend the granting, but not the withholding, of rights.

Speciesism

Speciesism is the concept of prejudicially placing humans above all other creatures. With its obvious parallels to racism and sexism, speciesism was expounded by Richard Ryder in a 1970 pamphlet (Waldau 2002). When lines drawn for sentience, reason, suffering, and moral obligation fail to separate all humans from all other animals, arguments against animal rights often fall back on the simple fact that we are human and other animals are not; therefore they do not deserve moral consideration (Frey 2002).

Species-based discrimination can extend beyond humans; many people, for example, arbitrarily place dogs and cats on a morally higher plane than deer, pigs, and sheep (Regan 2001; Fox 1986). Marc Bekoff rejects this idea of "higher" and "lower" species, and warns against making generalizations about entire species. He points out that many mice can outperform many chimpanzees in certain mimicry tasks, but that this does not make the mice superior, simply different from the chimpanzees (Bekoff 2002a). Analogously, humans cannot declare themselves superior to other species simply by outperforming the other species in self-described important tasks.

As Peter Singer (1975) points out, women and people of color have been denied basic rights on the mistaken assumption that they had inferior mental faculties. Singer proposes that humans cannot deny animals their natural lives because they cannot speak or compose symphonies.

The Inherent Abilities of Animals

Although many people value animals only as they relates to humans, many others readily ascribe importance to animals directly as inherently valuable sentient beings. Since ancient times, farmers, herders, pet owners, and other observers of animals have associated human feelings with various animal behaviors. As modern science provides further support for the notion of animal cognition, it becomes increasingly difficult to deny these claims and exclude animals from our moral world.

Pain

One very powerful experience humans share with many other animals is that of pain. Some philosophers count the ability to feel pain as the sole qualification for entry into the moral world (Rollin 1981), and it is apparent that at the very least, all vertebrates experience pain.

The concept of pain is distinct from nociception, which is simply a physiological response to any noxious (unpleasant or dangerous) stimulus. Pain includes a psychological state in addition to the physiological reaction (Block et al. 2004). Animals and humans display very similar physiological and neurological responses to painful stimuli, and this similarity in pain response is acknowledged by those companies and scientists who use animals to model human pain and develop pain-relieving drugs (Kest et al. 2000).

In addition to pain, it is possible that most vertebrates feel anxiety. Studies have shown that mammals, birds, lizards, frogs, turtles, and both bony and cartilaginous fishes all have benzodiazepene receptors. These receptor proteins are the targets of anti-anxiety drugs in humans (Nielsen et al. 1978;

Hebebrand et al. 1988). Amphibians mediate stress responses through corticosteroid hormones, just as humans do (Denver 1997). Furthermore, mice are used extensively in the research of human mood disorders (for examples, see Wei et al. 2004; Keck et al. 2004; Amico et al. 2004; Ray and Hansen 2004; Inoue et al. 2004). This is not to suggest that humans and other animals necessarily experience qualitatively similar anxiety, but since physiological links have been established, it is reasonable to assume there are psychological similarities.

Emotions

Beyond pain and anxiety, numerous scientific studies support the presence of such complex feelings as loss, generosity, and humor in animals (see Biller-Andorno 2002). And many find evidence for even more complex emotions into animal behavior. One of the more prominent people to assign human emotions to animals was Charles Darwin, who argued that "nature does not make jumps" and thus believed that human emotions could not have arisen spontaneously, but must have been developing throughout evolution (Darwin 1871).

> [T]he lower animals, like man, manifestly feel pleasure, pain, happiness, and misery.
>
> CHARLES DARWIN (1871)

Recent evidence suggests that not only do animals have emotions, but they go to some length to preserve positive emotions. For example, a study of an African troop of baboons documented a cultural shift toward non-violence in the troop after tuberculosis killed off the dominant, aggressive males. Over the course of the observation, during which there was complete turnover in the troop's male population, the non-violent culture remained intact, and all members of the troop showed lower than average stress levels. The results indicated that the females in the troop preferred the peaceful setting enough to teach their customs to incoming males (Sapolsky and Share 2004). Another recent study has shown that non-human animals experience uncertainty (Fiorillo et al. 2003), indicating that they realize their choices have consequences, and that these consequences can differentially beneficial or harmful.

Certainly those people who have studied apes find that these animals form friendships, practice a form of politics, and even have a sense of humor (Fossey 1983; Patterson and Linden 1985; Goodall 1988). Many people feel that because there is evidence that animals have emotions, and because these emotions are difficult to discern, people should err on the side of caution in preserving animal peace of mind, although others vehemently disagree with this position.

Animal Welfare in the Scripture of Major World Religions

Christianity

As an example of one religion encompassing many beliefs, Christianity has directed its followers both toward loving compassion for all living things, as well as the more traditional dominion over the natural world. However, the nature-empowering Christianity as preached by St. Francis of Assisi was never a major force (White 1967). The early Christian philosophers Augustine and Thomas Aquinas deemed animals subordinate to humans due to animals' lack of reason, and sanctioned the use of animals as means to an end. Thomas Aquinas declared that humans held no moral obligations toward non-human animals (DeGrazia 2002; Rollin 1981; Waldau 2002).

These interpretations of the Bible have changed little since the early days of Christianity, and in 1994 the Catholic Church printed a Catechism stating clearly that humans are a distinct group from other animals, and that even the minor interests of human beings justifiably override the major interests of animals (Waldau 2002). The Catechism also stated that people are allowed to perform medical research on animals within reason. These declarations reflect traditional interpretations of the Bible, in which God granted Adam dominion over all of His creation. The concept of man rising above a corrupt nature is much honored in the Christian tradition (de Sousa 1980). In this tradition, dominating nature and using it to benefit humans is a moral imperative. Traditional interpretations, however, also dictate respect for domesticated animals, although wild animals are viewed as dangerous (Waldau 2002).

More recent interpretations have noted that the Bible commands stewardship, respect for creation, and the preservation of nature for future generations (Waldau 2002). An gentler theological interpretation has been offered by Andrew Linzey in his book *Animal Theology* (1994). In Linzey's view, Jesus exemplified "lordship manifest as service," and his death represented a theological ideal of sacrificing the higher for the lower, not vice versa. Thus humans, being in the position of power, should make sacrifices to protect the wellbeing of other animals, which are in less of a position to assert authority and protect themselves (Linzey 1994). Linzey's view is in keeping with the New Testament depiction of Jesus as the Good Shepherd, caring for His flock and protecting them from harm.

Judaism

Judaism permits the use of animals by humans, but places a much greater emphasis on minimizing animal suffering than does Christianity (DeGrazia

2002; Wahrman 2002). Judaism's principle of *tsa'ar baalei chayim* (avoid causing pain to animals) allows animal use only if it fills a critical need. Thus hunting is allowed for food and clothing, but not for sport.

Similarly, prescriptions for the slaughter of animals (the Kosher rules) focus on ways to allow the animals as painless a death as possible. Jewish law allows animal suffering only if it will alleviate the suffering of a human. Thus, animal experimentation is allowable as long as the results lead to health benefits (Wahrman 2002).

Judaism has in Noah the archetypal "animal care person." An early Medieval rabbinic commentary (Tanhuma 2, 5) claims that Noah was God-like and righteous not merely because he housed all the animals, but also because he had studied them so well that he knew what and when to feed them. Concern is not enough; knowledge is also essential—a tenet that translates well to animal care in the modern laboratory. A modern Jewish commentary on *Genesis* holds that "Noah's wisdom is demonstrated in his painstaking detailed knowledge of the feeding schedules of every species within the ark … to be able to feed, one must have not simply sterling character, but a kind of curiosity, a kind of wisdom" (Zornberg 1995). The quotation from T. H. Huxley that opened the Preface of this book reflects just such a tradition: one must know the facts in order to do what is right. Zornberg reflects that "knowing the need" and being able to satisfy that need through one's knowledge is a characteristic of living a holy life.

Hebrew scriptures instruct people to treat their animals kindly, even to the point of requiring that animals, too, be allowed to rest on the Sabbath (*Exodus* 20:10). *Deuteronomy* 25:4 forbids the yoking together of animals of different strengths since it would cause hardship on the weaker animal. Even the Kosher restriction on eating meat with milk may have its origin in maintaining a respect for the animal and not to boil her young in her own milk (*Exodus* 34:26). One particularly striking example is the commandment to unload an animal staggering under its burden, even if it belongs to one's enemy (*Exodus* 23:5).

Rabbinic tradition is very adamant in protecting and respecting animals. Rabbi Judah declared that a person cannot eat until he gives food to his animals, and the Talmud proclaims that "the duty of relieving the suffering of beasts is a Biblical law." Several Jewish scholars were instrumental in movements that resulted in Britain's passage of the first modern laws against animal abuse. Foremost among them was Lewis Gompertz, co-founder of the British Society for the Prevention of Cruelty to Animals, whose book *Moral Enquiries on the Situation of Men and Brutes* (1824) galvanized support for animal protection measures (Levy 1996).

> God's compassion extends to all of Creation
>
> PSALM 145: 9

Islam

As in the other scriptural religions, Islam grants its followers dominion over other animals, although this rule is not unconditional. The Qur'an declares that animals possess intrinsic value, and people are to live in harmony with nature. Islamic law prohibits the destruction of an ant nest, even if the ants have harmed your property. According to tradition, all animals have higher mental faculties, invisible to humans because of human ignorance but evident in that all animals follow the instructions given to them by God (Masri 2004). Animals are also thought to have souls of varying types; according to one verse of the Qur'an "there is not an animal [*dabba*] in the earth, nor a creature flying on two wings but they are people [or communities] like you" [6:38] (quoted in Wescoat 1998).

Traditional Islam, like Judaism, accords all animals the right to protection, and domestic animals the right to water, shelter, and rest. Islamic scholars believe that all human disease is caused by humans and that animals need not suffer for this. The majority of animal research is generally seen by Muslims as nonessential and is not supported (Wescoat 1998; Foltz 2000; Masri 2004).

Buddhism

The Buddha taught his followers to live in harmony with nature, an edict in striking contrast to the Western traditions of domination and exploitation. Buddhism teaches compassion as a primary ethical value, and Buddhist monks are forbidden even to dig the soil, lest living beings be harmed. The high profile of animals in the Buddhist writings is seen as a sign of heightened respect for these animals. Whereas Scriptural theology usually postulates a large gap between humans and other animals, Buddhist traditions do not (see Katahara-Trish, 1961).

However, Waldau's *The Specter of Speciesism* (2002) points out that Buddhism is in fact a speciesist religion. The Vinaya states directly that all humans are superior to all other animals, although all non-human animals are seen as equal to each other. Reincarnation is a directional process leading toward the ultimate goal of birth as a human, who are the only beings capable of achieving enlightenment. Caught in the cycle of rebirth, all life is responsible for its own Karmic fate, and shame befalls any human who is reborn as a non-human. This shame may possibly be used as justification for the severe mistreatment of animals by Buddhists as documented by several scholars (Waldau 2002).

Buddhism sees the use of animals in research and education as permitted only if they serve a higher end. Even then, the animals should be treated with respect. A priest of the Soto Zen tradition stated that, "from a Buddhist point of view, anyone prepared to do this [experimentation] has to know and accept the karma of his actions. This would entail trying to do as little harm as possible, killing only if absolutely necessary, treating the being with tender respect, and making sure the knowledge is put to good use" (quoted in Kennaway 1980).

Animal Moral Standing and the Scientific Community

Scientists whose research requires them to work with and sometimes inflict pain and destroy laboratory animals must come to terms with these actions in their own minds. In this regard, three questions recur: what research activities actually harm the animals, and to what extent? and, is the use of animals in the research absolutely necessary?

What is harmful to an animal?

Does merely holding an animal captive harm it? What if it is in a very large or very small cage? For an animal taken from the wild, captivity can be extremely distressing. It has been placed in an unfamiliar environment in which it can no longer move around and perform its natural activities. If it is a social animal, such as a dog or a monkey, it is removed from other members of its social group. David DeGrazia notes the similarities to a prison—captivity as a punishment—for humans. After an extended period in captivity, however, the animal may resign itself to its new life and will no longer feel particularly distraught. If the animal is now protected from the dangers of the natural world, and if its life experience is only slightly dampened by captivity, being captive can be seen as beneficial, because of the protection from harms the animal would encounter in the wild. If, however, there is inherent value in "exercising one's natural capacities," then captivity harms the animal by preventing it from living its life as it would in the wild (DeGrazia 2002). This reasoning applies to laboratory animals since in most countries, it is still legal to obtain research animals from humane shelters and from the wild, rather than breeding them in captivity specifically for research (Orlans 2002).

The harm of captivity is much reduced in purpose-bred animals; not only are they likely to be kept in a familiar environment throughout their lives, but they cannot miss the wild life the never knew. People who have worked with wild and purpose-bred mice attest that purpose-bred strains of research mice are much less aggressive, and much less interested in leaving their cages, than their wild counterparts. In many instances, laboratory-bred animals also have traits, such as albinism or compromised immune function, that would make life in the wild nearly impossible (see Rader 2004).

The harms inflicted on research animals can be both psychological and physical. Physical restraint is often necessary and is traumatic for most animals. Frequent injections or blood sampling, injection of tumor cells or bac-

How Should Laboratory Animals Be Cared For?

As long as animals continue to populate laboratories—and that means for the foreseeable future—there are measures scientists and others can take that will improve conditions for the animals as well enhancing the accuracy of experiments. The most comprehensive regulatory guide to date is the "Guide for the Care and Use of Lab Animals" produced by the National Research Council for the U.S. National Institutes of Health (Institute of Laboratory Animal Resources Commission on Life Sciences 1996). The Guide covers many areas, from animal exercise to feeding to the hygiene of the animal care workers. However, it also leaves considerable regulatory power in the hands of Institutional Animal Care and Use Committees (IACUCs; www.iacuc.org). These committees are made up of at least five members and must include a veterinarian, a scientist with animal research experience, a non-scientist associated with the institution, and a person not associated with the institution (Gluck and DiPasquale 2002).

There have been concerns that IACUCs are inadequate and cannot be trusted to enforce appropriate guidelines at their respective institutions, and in support of the complaints a 1995 report of the Inspector General of the USDA found that 12 out of 26 IACUCs examined did not meet federal standards (Gluck and DiPasquale 2002). But there are those who come to the defense of these committees, arguing that they have greatly advanced animal care at most institutions. Bernard Rollin, one such defender, maintains that since the institution of IACUCs, scientists have become much more serious about using analgesia to diminish animals' pain, and that animal care is gradually moving out of the hands of the principal investigators running the laboratory and into the expert care of veterinarians and other specifically trained personnel. Rollin also asserts that many IACUCs have gone beyond what is required and have, for example, expanded their regulatory umbrellas to include invertebrates such as squid. Rollin admits that in many instances both IACUCs and the laws are inadequate, but advises not to underestimate the headway that has been achieved (Rollin 2002).

teria, brain cell destruction, and sleep, food, and water deprivation are some of the treatments to which laboratory animals have been subjected for medical research. In addition to these major interventions, studies have shown that small cages, unnatural living conditions, boredom, and lack of companionship can inflict greater harm on research animals than is necessary to conduct experiments (DeGrazia 2002).

The death of research animals is another issue. In the course of treatments or experimentation, animals are often sacrificed so that physiological or anatomical results can be studied. Most researchers feel that death, being the natural end to which all life must come, is in itself harmless, but there are other considerations. In the cases of some organisms, such as the mole-rat and certain bird species, removing one of a pair-bonded couple can cause anxiety in the remaining animal (DeGrazia 2002).

Is animal research necessary?

Most scientists and medical researchers would agree that without animal research many of the great breakthroughs of recent medical history would not have been possible. Nor would most of us want to be the ones to ingest pharmaceutical chemicals whose effects had not first been tested on animals. However, some among those who challenge animal research give it no credit at all in the development of human medicine, and claim that in some cases the use of animals has hindered medicine's progress.

Arguments for and against the use of animals in research generally involve the concept of homology. **Homology** is the idea that animals that evolved from a common ancestor develop and function in similar ways. The more closely we are are related to (i.e.,. the more recent the common ancestor) the other animal, the more our bodies are expected to share similar mechanics and anatomy. Thus, the discovery of insulin in humans came about via canine research and was possible because each dog has a pancreas that secretes insulin, just as humans do, and if you remove a dog's pancrease, it gets diabetes, just as a human would. Those who argue against animal experimentation focus on the *lack* of similarity, pointing out, for example, that different species metabolize biochemicals (such as drugs) differently, and that even different strains of mice metabolize differently.

To illustrate this point, Robert Garner (1993) mentions that arsenic is lethal to people but not to sheep, and iron oxide acts as a carcinogen in humans but not in hamsters, mice, or guinea pigs. Moreover, he states, morphine sedates people but stimulates cats, and penicillin is toxic to guinea pigs and hamsters but not people. These differences among species create odds in drug screening processes that might not be much different if drugs were screened at random, according to Daniel Von Hoff of the National Cancer Institute. Van Hoff estimates that if a drug shows anti-cancer activity in a mouse, there is about a 1 in 30 to 1 in 40 chance that it will show anti-cancer activity in humans (Reines 1986). If this is the case, some protest, how many potential anti-cancer drugs do we overlook because they are not active in mice?

Activists recount even more disturbing incidences in the history of animal research as evidence that, far worse than useless, the practice can be dangerous. The best-known cases are those of thalidomide and diethylstilbestrol (DES). Thalidomide was developed to combat the nausea and anxiety associated with pregnancy, and was determined to be safe in animal tests. Only after wide distribution of the drug was it seen to have severe teratogenic effects in humans. Similarly, the synthetic estrogen DES was found to be carcinogenic in humans after animal tests had deemed it safe as a drug to help maintain pregnancy (Reines 1986; Fox 1986).

Critics of the pharmaceutical industry assert that its focus is in the wrong place—profit and symptom treatment rather than human health and disease prevention (Garner 1993). Numerous animals are used and destroyed in order to develop so-called "me-too" drugs, which are slight variations on a common theme developed mainly for profit and to circumvent patent laws. Even in research areas of major activity and successes, such as cancer treatments, critics feel values are misplaced. For example, the majority of human cancer cases are at least partially the result of environmental exposure to elements such as radiation and chemical toxins (as in sun exposure and cigarette smoke). Many people feel that fewer funds should be devoted to the development of cancer drugs in animal models, and more should be spent on keeping people from getting cancer in the first place. And, they point out, many cancer therapies used today were actually first done in humans, either by accident or by researchers who had little faith in animal models (see Reines 1986).

Proponents of animal research such as Carl Cohen counter that even if all risk factors were eliminated from the environment—which could not possibly happen—many people would still get cancer, and we should not abandon the fight. Moreover, countless successful vaccines, cancer therapies, and surgical techniques were developed in animals before their transfer to human patients. Risk is inherent in drug development, and a few mishaps do not mean that animal testing is valueless. Animal studies weed out the most dangerous drugs, thereby preventing many human deaths. In addition, the recent ability of science to use transgenes (see Chapters 9 and 11) to create, for example, laboratory mice with human receptor proteins, makes the animal studies much more accurate in predicting a drug's behavior in humans (Cohen 2001).

Another supporter of animal studies, Michael Fox, lauds the importance of the scientific method. Experiments require fine control, he argues, which is possible only with laboratory animals whose diets, environments, etc. can be readily controlled and fine-tuned. Furthermore, he asserts, this control leads to repeatability and high confidence in statistical significance, which are difficult to attain in human studies (Fox 1986).

Despite species differences, there are many physiological, developmental, and immunological similarities among animals. Proponents of animal experimentation can point to such research as being critical in the discovery that insulin cures diabetes, that certain viruses cause cancers, that certain procedures will allow bones to heal, and that certain drugs were not allowed on the market because they caused cancer or disease in laboratory animals.

Alternatives to Animal Research: The "Three R's"

Increasingly, scientists are looking for alternatives to animal research. One model for this search was proposed by William Russell and Rex Burch in their 1959 book *The Principles of Humane Experimental Technique*, and is known as the "three R's": reduce, replace, and refine. New Zealand, the

Netherlands, Switzerland, Sweden, and the United Kingdom have adopted these three concepts in their legislation, and institutes such as the Johns Hopkins Center for Alternatives for Animal Testing (http://altweb.jhsph.edu/), are devoted to researching and implementing the principles outlined by Russell and Burch (Russell and Burch 1959; Robb 1988; Orlans 2002).

Reduce

The first R, "reduce," calls mainly for a re-evaluation of statistical methods. In many research studies, more animals are used than are required to establish statistical significance (that is, to verify that the results are truly meaningful and cannot be attributed to random chance). By increasing the power of the statistical test used (that is, by improving the mathematical formulae and computer algorithms used to compute the results), researchers could reduce the number of animals required still further. Testing of the "me-too" drugs mentioned earlier, as well as a majority of industrial toxicity testing, are also often claimed to be superfluous. Elimination of unnecessary testing could dramatically reduce the number of animals used in laboratories each year.

Replacement

"Replacement" champions the use of *in vitro* methods as well as computer models in place of animals in many fields. *In vitro* ("in glass") experiments use cells and tissues cultured in laboratory dishes rather than experimenting on living (*in vivo*) organisms. *In vitro* experiments can be done using human cells, thus testing drug effects on human tissues rather than animal models.* Researchers can observe whether a compound interacts with the targeted receptors in the human tissue, and in many cases, the toxicity of the compounds can be determined as well (Morton 2002; Zurlo et al. 2004). Tissue cultures can also be used in vaccine development and antibody production. The Netherlands, Switzerland, and the United Kingdom recently banned the ascites method, in which antibodies are produced by injecting tumor cells into a mouse's abdomen. This procedure causes severe pain and distress in mice. In its place British researchers use *in vitro* methods, in which antibodies are made in plastic tubes. The ascites method is still used in the United States.

Cancer research can also make use of *in vitro* methods. Methods for culturing human tumor cells have been available since the 1970s, and in 1979,

---

In vitro methods have replaced a great deal of the animal testing formerly done by the cosmetics industry, although some products (most notably hair dyes) still require a certain level of animal testing.

Daniel Von Hoff and Sidney Salmon demonstrated independently that cultured cells can predict with 96 to 100 percent accuracy that a given anti-cancer drug will *not* work in clinical trials (Reines 1986). Subsequent studies have shown that *in vitro* assays may also prove helpful in determining whether a specific drug *will* work (Link et al. 1996; Ness et al. 2002). Likewise, suspected carcinogens can be identified through *in vitro* tests (such as the Ames test , which has been widely used since 1974) that test the ability of varioius substances to cause mutation in bacteria (Reines 1986). Epidemiological studies (statistical compilations) of human populations can also offer insight into the causes and possible prevention of cancer (DeGrazia 2002).

Technological advances such as the synthetic skin Corrositex®, which can be used for toxicity testing, and computer models have also proven useful in replacing lab animals. As we learn more about protein structure and can model protein interaction on computers, such modeling systems hold promise in both toxicity testing and drug screening (Reines 1986; Fox 1986; Robb 1998).

Refinement

The final R, "refinement," seeks to fine-tune experiments to minimize the amount of stress placed on animals that remain in laboratories after the first two R's have been applied. The benefits of such techniques extend beyond kindness to animals, and result in more reliable data, and therefore reduced cost and effort. Since Russell and Burch first articulated this most complicated and murky of the three R's, many techniques and regulations have been developed to improve the living conditions of lab animals.

Housing is one potential source of stress for animals, and it should be as natural and as comfortable as possible. Even seemingly insignificant factors, such as lighting can mean the difference between healthy animals and blind, vitamin D-deficient animals with disrupted neuroendocrine function (Brainard 1989; Williams 1989). In addition, a recent study has shown that socially reared rats are less fearful and show lower stress levels than those raised without companions (Molina-Hernandez and Tellez-Alcantara 2004). David DeGrazia (2002) has suggested that the animals should be able to make choices—for example, they could be offered two living environments that they can choose between.

Before starting *in vivo* work on animals, researchers are encouraged to perform as much *in vitro* work as possible. When *in vivo* work is necessary, small pilot studies can confirm or allay fears about how the animals will react to the study. A researcher should use as minimally invasive a procedure as possible. Increasingly sensitive modern biochemical tests can detect compounds in ever-smaller tissue samples, and assays based on urine output rather than blood tests can eliminate recurrent blood drawings (Hiebert et al. 2000).

To improve the track records of Institutional Animal Care and Use Committees (IACUCs; see p. 255) and other regulatory bodies, scientists and ethicists have suggested an "experimental checklist" that holds zero animal suffering as its ideal (Porter 1989; Morton 2002). Before beginning an animal study, the researcher should investigate (1) whether the experiment had been performed before; (2) possible alternatives to the use of sentient animals; and (3) the importance of the experiment in light of the distress that the animals might endure (Morton 2002). To aid researchers in this final task, Nikola Biller-Andorno (2002) has suggested an empathy-based approach to understanding animal suffering. She advises looking at the experiment from the animal's point of view and trying to see how it would perceive each aspect of the experiment. Although many scientists dismiss this as impossible, Biller-Andorno maintains that such an approach is practical rather than emotional.

If the researcher decides to proceed with the experiment, he or she should explore different statistical methods of reducing the number of animals to the absolute minimum number necessary. The researcher must also take great care in choosing an experimental subject. The checklist would force the experimenter to consider different species and strains of animals and select the most appropriate for the study (Morton 2002). The decision should include analysis of the physiological as well as behavioral components of the animals' nature. The experimenter should also take into account the different personalities of the animals and notice which caretakers the animals react most favorably to, as any additional stress to the animals will add an extra dimension of variability to the study. The overall effect of such a checklist, say its promoters, would be improved research results as well as less animal suffering (Guttman et al. 1988).

The other "Rs": Remembrance and respect

Susan Ilif has suggested that Oriental Buddhist and Shinto traditions have added a "fourth R" to the three classic R's guiding animal welfare: remembrance (Ilif 2002). Indeed, in Japan, there are memorial stones at over 170 institutions to commemorate the animals whose lives were sacrificed for the sake of human welfare (Kast 1994). In the spring of 1993, a "soul-consoling stone" was set up on the lawn of the Animal Research Institute of the Chinese Academy of Medical Science in Beijing (Holden 2004; Figure 15.1). This stone is a tribute to the thousands of animals sacrificed as researchers raced to develop a vaccine against SARS (Severe Acute Respiratory Syndrome) in late 1992. Even though the Chinese Communist regime suppresses organized religion, the notion that "human beings or animals, we are all Nature's creatures" survives (Qin, quoted in Holden, 2004).

In addition to commemorative stones, several Asian institutions hold services where they pay respects to the animals used in their research.

FIGURE 15.1   The "soul-consoling stone" outside the Animal Research Institute of the Chinese Academy of Medical Science in Beijing. The stone was erected in memory of the thousands of animals sacrificed in the development a vaccine against SARS. (Photograph © China Features, Beijing.)

These include the *ireiai* and *kanshasai* ceremonies (comforting and giving thanks) in Japanese university primate centers (Asquith 1983). Similarly, the Food and Drug Administration of South Korea has an annual event to acknowledge the contributions of their research animals (Choi 1988).

Donna Haraway, a philosopher with concern for both animals and biology, points out that part of being human is existing in and preserving a network of relationships with other species (Haraway 2003). Refining our experimentation procedures and according respect to our research animals may help to alleviate stress for scientists as well as for the animals. Killing animals is not easy for many investigators, and acknowledging the role of animals in our research may help scientists deal with the fact that new understanding can sometimes be obtained only by sacrificing animal lives. By according these animals all possible respect both during their lives and after their deaths, we need not sever the bonds between us and other animals.

Glossary

A

acrosome   A membrane-enclosed sac at the tip of the sperm cell, containing enzymes that digestproteins and complex sugars. This sac releases these enzymes to digest a hole through the outer coverings of the egg.

adenine (A)   One of the nitrogen-containing bases that make up the nucleic acids (DNA and RNA). See also **base; purine**.

allantois   A sac-like extraembryonic (located outside the embryo) membrane that contains the waste products generated by the embryo.

amnion   The water-filled sac in which the embryo develops.

amniocentesis   Method of making a prenatal diagnosis by taking a sample of amnionic fluid at the fourth or fifth month of gestation. Fetal cells in the fluid can be grown and then analyzed for the presence or absence of certain chromosomes, genes, or enzymes.

amnionic sac   The sac that surrounds the embryo with amnionic fluid, protects it from drying out, and cushions it against impacts.

ampullary region   The region of the oviduct (Fallopian tube) where fertilization takes place. It is located close to the ovary.

anencephaly   A neural tube defect characterized by a failure to close the anterior (head) portion of the neural tube. The brain fails to form in such embryos.

anti-Müllerian duct hormone (AMH)   Hormone secreted by the Sertoli cells in the developing testes which causes the degeneration of the Müllerian duct. This causes males to lack uteruses and oviducts. Also called Müllerian inhibitory substance (MIS).

assisted hatching   A technique used as a part of in vitro fertilization to ensure that an embryo is able to hatch from the zona pellucida in time to adhere to the uterus. The technique creates a microscopic hole in the zona pellucida prior to inserting the embryo into the uterus.

assisted reproductive technologies (ART)   Techniques that enhance the probability of fertilization by directly manipulating the oocyte outside the woman's body, then implanting the embryo in the woman's uterus. These techniques were designed to allow previously infertile couples to conceive a child.

autosomes   Any chromosome other than a sex chromosome. In humans, the 22 numbered chromosomes. Contrast with **sex chromosome**.

azoospermia   Medical condition in which a male has an extremely low number of viable sperm.

B

base   In nucleic acids (DNA and RNA), the purine (adenine or guanine)or pyrimidine (cytosine, thymine, or uracil) that is attached to each sugar in the backbone.

bipotential stage   Period in gonad development when they have neither female nor male characteristics. In humans, the bipotential stage appears during week 4 of development and lasts until about week 7.

birth defect A physiological (relating to function) or structural abnormality present at or before birth as a result of a gene defect, developmental irregularity or injury to the conceptus. Also called a congenital anomaly.

blastocyst Stage in early mammalian development when the conceptus has an inner cell mass surrounded by the trophoblast (outer layer of cells); the stage which implantys into the uterus and from which stem cells can be derived.

blastomeres The cells produced by the first few divisions (cleavages) of a fertilized egg.

blastula An early stage in animal embryology; in many species, a hollow sphere of cells surrounding a central cavity. In mammals, the blastocyst is the stage equivalent to the blastula of most animals.

C

capacitation The final stages of sperm maturation that occur inside the female reproductive tract, and which allow the sperm to bind to the zona pellucida and fertilize the egg. Sperm that are not capacitated are "caught up" in the cumulus and never reach the egg.

cell-cell interactions Communication that occurs between cells and coordinates cell activity across the body. Such communication is necessary for stem cell differentiation. Cell-cell interactions are frequently mediated through paracrine factors.

cell membrane The layer of fats and proteins that encloses a cell. The cell membrane regulates the flow of certain ions during fertilization, has the receptors for paracrine factors,k and allows the cell to interact with other cells and with its environment.

centriole A cellular structure that produces protein fibers that help to organize the chromosomes during cell division.

centromere The region where sister chromatids join. In the first meiotic division, the centromere holds the chromatids together; and in the second meiotic division, the centromere splits, creating haploid cells.

chorion The embryonic portion of the placenta in mammals that comes in contact with the mother's blood vessels such that oxygen and nutrients may pass through to the developing embryo or fetus. It is derived from the trophoblast of the blastocyst.

chorion biopsy Method of making a prenatal diagnosis by taking a sample of the placenta. Fetal cells can be grown from the sample and then analyzed for the presence or absence of certain chromosomes, genes, or enzymes.

chromatid One of the pair of chromosomes held together by the centromere during the first meiotic division. Each chromatid is a double helix of DNA. The pair of chromatids representing DNA that has replicated prior to meiosis are called sister chromatids.

chromosome The structure composed of DNA and proteins, found in the nucleus, that bears most of the genetic information of the animal cell.

cilia Hairlike projections on the cell surface whose actions generate or direct movement. Within the oviduct, cilia move the early embryo toward the uterus, where it implants.

cleavage The first several cellular divisions of the fertilized egg (zygote) of an animal.

clone Genetically identical cells or organisms produced from a common ancestor by asexual means. Cloning is easy in plants using a cutting (root, stalk or runner) from the parent. Vertebrate cloning involves transferring nuclei from one animal into enucleated oocytes and having them develop by procedures other than sperm entry

cloning (1) For animals: The process of transferring a new nucleus into the "empty" egg by opening a donor cell and transferring its nucleus into the oocyte through a micropipette. Animals produced in this manner are genetically identical. Also called **somatic cell nuclear transfer (SCNT)**. (2) For DNA: isolating a region of DNA and putting it into a bacterium so that it will be replicated millions of times.

codon Three nucleotides (one of 64 possible combinations composed of adenine, cytosine, guanine, or thymine) in messenger RNA that direct the placement of a particular amino acid into a polypeptide chain. May also be called a **nucleotide triplet**.

committed stem cell A stem cell (since it can divide to produce both a more differentiated type of cell as well as another stem cell of the same type), but the type of cell will always be the same (i.e. a blood stem cell only produces blood cells). Sometimes called a **lineage-restricted stem cell**.

compaction   Process whereby blastomeres of mammalian embryos huddle together forming a compact ball of cells bound by tight junctions between the outermost cells of the ball. This compact ball will form the morula.

complementary base pairing   The A-T (or A-U), T-A (or U-A), C-G, and G-C pairing of bases in double-stranded DNA; in transcription, the pairing between DNA and mRNA.

conception   Term used to describe the process of fertilization in humans.

conceptus   The products of conception, including the zygote, the embryo, the embryonic sacs and the fetus.

congenital anomaly   An anatomical abnormality, as a result of a gene defect, developmental irregularity or injury to the conceptus, seen at the time of birth. Also called a birth defect.

cortex   The outer layer. In a cell, the thin layer of gel-like cytoplasm lying immediately beneath the egg's cell membrane.

cortical granules   Spheres located in the egg's outer cytoplasm which contain protein-digesting enzymes. The cortical granules release their contents after fertilization.

crossing over   An exchange of chromosomal material that occurs during meiosis and results in a unique single chromosome containing a mix of genes from both parents. Crossing over creates trillions of unique chromosomes possibilities.

cryopreservation   The process whereby an embryo is frozen, and then thawed before implantation into the uterus. Sperm and embryos may be frozen, stored, and then thawed before use; unfertilized eggs cannot.

cumulus   The outer "cloud" of cells surrounding the ovulated oocyte. It is made up of the ovarian follicle cells that were nurturing the egg at the time of its release from the ovary

cytoplasm   The cellular material, excluding the nucleus. It contains ribosomes, mitochondria, and is rich in proteins.

cytosine (C)   One of the nitrogen-containing bases that make up the nucleic acids (DNA and RNA). See also **base; pyrimidine**.

D

decidua   The expanded outer lining the uterus which becomes the maternal portion of the placenta during pregnancy.

development   The process of progressive change within an animal that generates cellular diversity and ensures the continuity of life from one generation to the next. Development continues throughout the life of the animal.

developmental biology   The branch of biomedical science that studies the development and transformation of embryos. The more modern name for the field of **embryology**.

differential gene activation   Activation or expression of certain genes only in certain cell types (i.e., hemoglobin genes are active only in those cells that will become red blood cells) or at certain times in the lifespan of the cell.

differential RNA processing   Method of preparing mRNA by cutting out a particular intron, and leaving a particular exon, such that the same gene can generate many different mRNA sequences coding for different proteins by removing or retaining different introns and exons. Someimes called differential RNA splicing.

differentiation   Process in embryonic growth whereby cells become different from each other and from their progenitors. Thus, originally similar cells follow different developmental pathways, resulting in a diversity of cells types specialized for different purposes (i.e., red blood cells, muscle cells, neurons).

diploid   Having each chromosome represented twice (in humans, one copy each of 23 chromosomes from the mother, and one copy each from the father) within a cell. The human diploid chromosome number is 46. Contrast with **haploid**.

disruptions   Congenital anomalies caused by exposure to external agents, such as certain chemicals, viruses, radiation, or high fevers. Contrast with **malformations**.

dizygotic   A classification of human twins characterized by the fertilization of two separate eggs to form two genetically different embryos; fraternal twins.

dorsal   Pertaining to the back or upper surface of an animal. Compare with **ventral**.

DNA (deoxyribonucleic acid)   The fundamental hereditary material of all living organisms, made up of paired molecules called nucleotides that hold the molecular code for the 20 amino acids. In animals, DNA is found primarily in the chromosomes within the nucleus. It is a nucleic acid using deoxyribose sugar rather than ribose sugar.

Down syndrome   A malformation (type of congenital abnormality) caused by an extra copy of chromosome 21 in each cell. It is characterized by mental retardation, the absence of a nasal bone, heart defects, a slanting of the eye, and often the closure of the intestine.

duality   The concept that the human body and the human soul are separate entities.

E

ectoderm   The outermost layer of cells in the embryo that generates the surface layer (epidermis) of the skin and also forms the brain and nervous system.

ectopic pregnancy   A condition whereby the blastocyst implants into the oviduct wall, as opposed to the uterus. Also called a "tubal" pregnancy.

electroencephalogram (EEG)   A graphic recording of the electrical potentials from the brain. May be used to monitor brain activity in fetuses or adults.

embryo   Term used to describe the conceptus during the first 8 weeks of gestation, the time when cells differentiate and organ systems begin to form.

embryogenesis   The processes between fertilization and birth that comprises conceptus development.

embryology   The branch of biomedical science that studies the development and transformation of embryos. Also called **developmental biology**.

embryonic stem cells (ES cells)   Stem cells that are derived from the inner cell mass of a blastocyst or from primordial germ cells of the fetus.. Such cells may be grown in a controlled setting (i.e. a flask) such that they remain undifferentiated and continue to divide. The cells can be directed to form the precursor cells that give rise to blood cells, nerve cells, and other cell types.

endoderm   The innermost layer of cells in the embryo which gives rise to the lining of the digestive tube and its associated organs (including the lungs, esophagus, stomach, and gut).

endometrial cell   The cells lining the inside of the uterus which catch the blastocyst.

enhancer   Sequence of the gene (on the chromosome) that regulates gene expression. Proteins (transcription factors) will bind to the enhancer, turning the gene on or off by signaling the promoter whether it should be active, and how much mRNA product it should synthesize.

enucleate   To be without the nucleus. The removal of the nucleus to create an enucleate egg is the first step of the cloning process.

ensoulment   In many philosophies and religions, a point at which a human soul is believed to enter the developing body. There is no consensus among different beliefs as to the exact time frame; the Aristotelian belief that ensoulment occurred around day 40 of development was widely held throughout much of Western history.

essentialism   A philosophical belief in the true essence of things; in other words, that there are invariable and fixed properties which define the "whatness" of a given entity. Used in some theologies to define the zygote as a human being.

estrogen   Hormone produced by the ovaries in female embryos (XX embryos) that enables the Müllerian duct to develop into the uterus, oviducts, and upper end of the vagina. Estrogen also stimulates the onset menstruation and assists in the regulation of ovulation. It is also used in men for sperm concentration.

exons   The regions of the genes that encode the amino acids of the protein. The regions of the gene that are represented in the mRNA. Contrast with **introns**.

extraembryonic   A structure that is located outside the embryo.

F

Fallopian tube   The tube in placental mammals that transports the egg from the ovary to the uterus. Sperm must also swim up this tube. Also called the oviduct.

fertilization   A multistep process involving the fusion of sperm and egg to produce a zygote and the joining of the haploid chromosomes from both parents to produce a new diploid genome.

fetus   Term used to describe the human conceptus during the final 30 weeks of gestation, the time when organ systems mature and the organism increases size.

fimbriae   The fingerlike projections of the oviduct that recognize the egg and transfer the egg from the ovary to the oviduct.

flagellum   Long, whiplike tail that propels cells. The flagellum is used by human sperm to travel to the oviduct.

folic acid   Vitamin that is crucial for proper neural tube development. Also called folate or vitamin B9.

G

gain-of-function mutation   An alteration of a gene such that the resulting protein has a novel ability not present in the wild-type (normal) gene. Contrast with **loss-of-function mutation**.

gametes   The mature sex cell: the egg (in female) or the sperm (in males). Compare with **somatic cells**.

gamete intrafallopian transfer (GIFT)   A variation of IVF (in vitro fertilization) that allows fertilization to occur naturally within the female partner's oviduct, instead of in the laboratory. In this procedure, eggs and sperm are mixed together in the woman's oviduct.

gastrula   An embryo forming the characteristic three cell layers (ectoderm, endoderm, and mesoderm) which will give rise to all of the major tissue systems of the adult animal.

gastrulation   The coordinated series of cell movements and interactions through which the three germ layers arise. After gastrulation, cell fates become specified.

genital ridges   Two pairs of gonadal rudiments, one on each side of the lower abdomen, composed of an internal compartment of loose cells and an external compartment of tightly connected cells called the sex cords.

genome   The complete set of an individual's genes. Contains the instructions for the formation and growth of the individual.

genotype   An exact description of the genetic constitution of an individual, either with respect to a single trait or with respect to a larger set of traits, needed by the embryo to grow in particular ways.

germ cell   A gamete or the cellular precursor of the egg (female) or sperm (male). Contrast with **somatic cell**.

germline   The lineage of germ cells from the earliest embryonic cells through the gametes.

germline gene therapy   The insertion of a wild-type (normally functioning) gene that would substitute for a defective gene into an individual's germline. Sometimes called **inheritable genetic modification (IGM)**. Contrast with **somatic cell gene therapy**.

germ layers   The three embryonic tissue layers (ectoderm, mesoderm, endoderm) formed during gastrulation. (From the same root as "germinate.")

gestation period   The time between fertilization and birth when the embryo and fetus develop within the mother; 38 weeks for humans.

gonad   An organ that houses the sex cells and in which they mature: either an ovary (female gonad) or testis (male gonad).

growth factors   A group of proteins (used in cell-cell interactions) that are secreted by one group of cells to trigger the normal growth and differentiation of neighboring cells within the embryo. Each growth factor acts only on certain target cells That have the receptor for that particular factor. Also called **paracrine factors**.

guanine (G)   One of the nitrogen-containing bases that make up the nucleic acids (DNA and RNA). See also **base; purine**.

H

haploid   Having a chromosome complement consisting of just one copy of each chromosome. Gametes, produced by meiosis, have a haploid nucleus, containing 23 chromosomes in humans. Contrast with **diploid**.

hormone   A substance produced by one set of cells which enters the blood stream and influences the development of another set of cells. The receiving cells have receptors that will bind that hormone. They are also called endocrine factors. Compare with **paracrine factor**.

human chorionic gonadotropin (hCG)   The hormone secreted by the trophoblast cells in the embryo that triggers progesterone production by the ovaries and helps maintain a hospitable environment for the embryo within the uterus. This hormone is the basis for pregnancy tests.

hydatidiform mole   A growth that resembles a mass of placental tissue and lacks an embryo. A hydatidiform mole occurs in about one out of every 1500 pregnancies and many of them have been shown to arise from a haploid sperm fertilizing an egg in which the female nucleus is absent.

I

induction The specifying of cell fate, whereby the cells are instructed (usually by paracrine factors) by their neighbors. The most common way that cell fates are determined in vertebrate embryos.

inheritable genetic modification (IGM) See **germline gene therapy**.

inner cell mass (ICM) Formed from the innermost cells of the morula (16-cell stage in embryo development), these cells give rise to the embryo and the yolk sac. When harvested and cultured, they can become embryonic stem cells.

insulator element Regulatory region of the gene that prevent the enhancer from stimulating gene activity in neighboring (i.e., not the target sequence) genes.

intracytoplasmic sperm injection (ICSI) A variation of IVF (in vitro fertilization) whereby a single sperm is injected directly into the cytoplasm of an egg. This method is helpful for men who have low sperm production.

introns Portion of a DNA molecule that, because of RNA splicing, is not involved in coding for part of a protein; It is not present in the messenger RNA. Contrast with **exons**.

ions Electrically charged molecules; an atom or group of atoms with electrons added or removed, giving the particle a negative or positive electrical charge.

in vitro fertilization (IVF) ART (assisted reproductive technologies) procedure that involves retrieving eggs (from a female) and sperm (from a male) and placing these gametes together in a petri dish (laboratory container) to enhance fertilization. After the fertilized eggs have begun dividing, the blastocyst-stage embryos are transferred into the female's uterus, where implantation and embryo development can occur as in a normal pregnancy.

K

karyotype The number, forms, and types of chromosomes in a cell; the karyotype of a human includes 46 chromosomes.

L

labor The process of birth.

Leydig cells The male gonadal cells derived from the loose tissue around the seminiferous tubules, They secrete the hormone testosterone.

lineage-restricted stem cells Stem cells that can form a restricted group of cell types (i.e. mesodermal stem cells may only form mesodermal organs, ectodermal stem cells may only form epidermal or neural systems and endodermal stem cells may only form the gut or lung). Sometimes called committed stem cells. Contrast with **totipotent stem cells** and **pluripotent stem cells**.

loss-of-function mutation An alteration of a gene such that it no longer functions. It either fails to encode a protein or encodes a non-functional protein. Contrast with **gain-of-function mutation**.

M

major histocompatibility complex (MHC) A complex of linked genes that produce proteins of the immune system. These proteins are found on the membrane surfaces of the body's nucleated cells and they mark the cells as being "self". These proteins are needed for the body's defense against any material.

malformations Congenital abnormalities caused by genetic events, such as gene mutation or an abnormal chromosome number. Contrast with **disruptions**.

meiosis Division of a diploid germ cell to produce four haploid cells (egg in females and sperm in males). The process consists of two successive nuclear divisions with only one cycle of chromosome replication. Meiosis is the process by which all gametes are formed. Contrast with **mitosis**.

methylation The addition of a methyl group ($—CH_3$) to a molecule. Extensive methylation of cytosine (C) in DNA, which prevents transcription factors and RNA polymerase from binding to the promoter, is correlated with reduced transcription.

mesoderm The layer of cells sandwiched between the ectoderm and the endoderm which generate the blood, heart, kidneys, gonads, bones, muscles, and connective tissues (e.g., ligaments and cartilage).

mitochondria The cellular organelle responsible for producing energy needed to maintain cell functions.

mitosis Process of cellular division resulting in two cells with identical chromosomal content to that of the parent cell. Mitosis is the process by which all somatic cells of the body divide and multiply. Contrast with **meiosis**.

monozygotic   A classification of human twins characterized by the splitting of a single zygote or early embryo to form two embryos with an identical genotype (chromosomal makeup); identical twins.

morphogenesis   The creation of ordered physical form over the course of embryonic development; includes organogenesis, the formation and growth of the body's organs, tissues, and organ systems.

morula   Solid cluster of blastomeres, consisting of a small group of internal cells surrounded by a larger group of external cells, formed at an early stage in development from the compacted embryo.

mRNA (messenger RNA)   Molecule that is a transcript of one of the strands of DNA (composed of exons) and that carries the information (as a sequence of codons) for the synthesis of a protein.

Müllerian duct   Duct system seen in early embryos that can differentiate into the oviducts, uterus, cervix, and upper vagina in female mammals.

mutation   A detectable, heritable change in the genetic material not caused by recombination; an alteration in the sequence of nucleotide bases encoded in the DNA or mRNA.

N

neurulation   Process whereby the neural tube is formed.

neural tube   An early stage in the development of the vertebrate nervous system consisting of a hollow tube that will become the brain and the spinal cord of the animal.

nucleotide   The basic chemical unit in a nucleic acid (DNA or RNA). A nucleotide in RNA consists of one of four nitrogenous bases (adenine, cytosine, guanine, and uracil) linked to ribose, which in turn is linked to phosphate. In DNA, deoxyribose is present instead of ribose and thymine replaces uracil.

nucleotide triplets   Three nucleotides (one of 64 possible combinations composed of adenine, cytosine, guanine, and thymine) in messenger RNA that direct the placement of a particular amino acid into a polypeptide chain. More commonly called **codons**.

O

oocyte   The cell that gives rise to the eggs in animals; an immature egg.

organogenesis   The formation of organs and organ systems by the three germ layers (ectoderm, endoderm and mesoderm) during embryo development.

ovary   The female reproductive organ that produces eggs.

oviduct   The tube in that transports the egg from the ovary to the uterus. Also called the Fallopian tube.

ovum   The egg, the female sex cell.

P

paracrine factors   Chemicals used in cell-cell interactions that are usually proteins and that are secreted by one cell type and are received by neighboring cells. Such signals are used locally to contact only cells in close contact with the sender. Often called **growth factors**.

period of susceptibility   The period of time during the conceptus' development when the embryo or fetus is vulnerable to damage by a particular teratogen.

phenotype   The observable traits (e.g., eye and hair color; height; level of cholesterol production) of an individual. The phenotype is the result of interaction between the genotype and environmental factors.

placenta   The organ found in most mammals that provides for the nourishment and respiration of the fetus, and elimination of the fetal waste products. It is made partly from the trophoblast of the embryo and partially from the uterus of the mother.

pluripotent   The ability of a cell to form all the cell types of the body, except for the trophoblast). This ability is thought to be a property of the inner cell mass blastomeres and of embryonic stem cells. Contrast with **totipotent**.

pluripotent stem cells   Stem cells that are capable of forming any cell of the embryo, but have lost their ability to form trophoblast cells. The embryo cells of the inner mass are called pluripotent stem cells. Contrast with **totipotent stem cells**.

polar body   A nonfunctional nucleus produced by meiosis, accompanied by very little cytoplasm. Meiosis that produces a single mammalian egg also produces three polar bodies.

polyspermy   The process by which multiple sperm enter the egg. The result is a nucleus with too many chromosome sets, and an unviable zygote.

preimplantation genetic diagnosis   The screening of embryonic cells before the embryo is inserted into the uterus for implantation. The procedure may be performed on embryos created by in vitro fertilization and is used to prevent the development of a baby that will display a genetic disorder.

prenatal diagnosis   The diagnosis of a genetic disease, often by chorion biopsy or amniocentesis, before the baby is born.

pre-mRNA   Initial gene transcript containing the intervening gene sequences (introns) that must be removed to produce the functional mRNA.

primary sex determination   The determination of whether an individual's gonads (sex organs) become testes or ovaries. In humans, the sex chromosomes dictate gonad development for females (two X chromosomes) and males (one X and one Y chromosome). Contrast with **secondary sex determination**.

progenitor cells   Cells capable of cell division, that have not differentiated, but whose progeny will differentiate into a specific cell type. Contrast with **stem cells**.

progesterone   A female sex hormone in humans, and vertebrates, that maintains pregnancy.

promoter   The region of DNA containing a specific sequence that binds the RNA polymerase, allowing it to unwind the double helix and initiate the synthesis of mRNA.

protein   One of the most fundamental building substances of living organisms. A long-chain of amino acids. Structural proteins such as collagen and elastin provide support for tissues and organs; enzymatic proteins such as alcohol carry out the work of metabolism that keeps us alive; and transport proteins such as hemoglobin deliver small molecules and atoms from one part of the body to another.

protein pharmaceuticals   Proteins that are normally found in the body but that may be lacking or defective in some individuals and thus need to be isolated and administered as drugs. Important protein drugs include insulin, protease inhibitor, and blood clotting factors;

these proteins are all difficult to manufacture biochemically.

proteome   The total number of proteins that the genome is capable of synthesizing. Because of alternate splicing of pre-mRNA, the number of proteins that can be made is usually much larger than the number of protein-coding genes present in the organism's genome.

purine   Adenine and guanine. The larger of the two molecular types of nitrogenous bases found in nucleic acids (DNA and RNA), each pairs with a corresponding **pyrimidine**.

pyrimidine   Cytosine, thymine (DNA), and uracil (RNA). The smaller of the two molecular types of nitrogenous bases in nucleic acids (DNA and RNA), each pairs with a corresponding **purine**.

Q

quickening   The point at which the fetus can be felt moving within the uterus; occurs at approximately 120 days (4 months) gestation.

R

receptor   A protein, in the cell membrane or within the cell (cytoplasm) to which a specific signaling agent (protein, hormone or other chemical) from another cell binds. Used for **cell-cell interactions**. Receptors can bind paracrine factors or hormones.

reciprocal induction   Process whereby neighboring cells interact through their paracrine factors such that they both influence the differentiation and morphogenesis of each other. in this way, adjacent cells can interact with one another to create organs.

regulation   The ability of the embryo to form a complete individual even after cells are removed from it; the plasticity of embryonic cells to develop in accordance with their new location if transplanted into a different region of the embryo

reproduction   Process of creating of a new organism. In animals this is accomplished by the fusion of two gametes (the female egg and male sperm) in what is called sexual reproduction.

ribosomes   A cluster of particular RNAs and proteins in the cytoplasm which is the site of protein synthesis. Messenger RNAs bind to the ribosome and the ribosome makes certain that the appropriate amino acid is brought to the particular codons and joined into a protein.

RNA (ribonucleic acid)   An often single-stranded nucleic acid whose nucleotides use ribose, rather than deoxyribose as in DNA, and in which the base uracil replaces thymine found in DNA. (See **mRNA**.)

RNA polymerase complex   An enzyme that catalyzes the formation of RNA from a DNA template by opening up the double strand of DNA and binding to one of the strands (the template). It then proceeds along that strand, using nucleotides present in the nucleus to create an RNA strand that is complementary to the template DNA strand

S

secondary sex determination   The specification of the sexual phenotype through hormones. These include the form of the external genitalia, distributions of body fat and body hair, differences in muscle mass, voice tone and pelvic bone structure, and the male or female duct systems. Contrast with **primary sex determination**.

Sertoli cells   The somatic cells, derived from the internal male sex cords, in the seminiferous tubules that nurture the developing sperm in males. In the embryo, they secrete anti-Müllerian duct hormone.

sex chromosomes   In animals, the chromosomes involved in sex determination. These chromosomes are not referred to by number, but are designated X and Y. For primary sex determination, males are XY and females are XX.

sex cords   The cells that will form the tissues that hold and nurture the germ cells; if the gonadal cells have the XY genotype, the sex cords grow inwardly and the germ cells within them become sperm; if the gonadal cells have the XX genotype, the sex cords develop in the periphery of the gonad, and the germ cells migrating into them will become eggs.

somatic cell   Cells of the body, excluding egg or sperm. Includes tissues and organs. Contrast with **gametes** and **germ cells**.

somatic cell gene therapy   The insertion of a wild-type (normally functioning) gene that would be activated at the appropriate times and places into somatic cells (of a single individual). Such gene therapy has the potential to cure a genetic disease (i.e., immunodeficiency or cystic fibrosis). Contrast with **germline gene therapy**.

somatic cell nuclear transfer (SCNT)   The process of transferring a new nucleus into the "empty" egg by opening a donor cell and transferring its nucleus into the oocyte through a micropipette. Animals produced in this manner are genetically identical to the animal who donated the nuclei. More commonly called **cloning**.

sperm selection   A method for separating sperm carrying an X or a Y chromosome by using the difference in sizes of these chromosomes. The process allows couples to chose the sex of their offspring.

spermatocyte   The immature haploid sperm cell necessary for male reproduction.

spina bifida   A neural tube defect characterized by the failure to close the posterior (toward the bottom) end of the neural tube.

stem cells   In animals, a cell that can divide so that one of the products of division is a more differentiated cell type (such as an immature sperm cell or a red blood cell), while the other product is another stem cell just like the original cell, which can continue to divide for an indefinite period of time. See also **embryonic stem cells**, **totipotent stem cells** and **pluripotent stem cells**.

surfactant   A type of protein secreted by cells of the alveoli of the lungs, in mature mammals, that when secreted by the fetus starts the protein cascade responsible for labor. Surfactant refers to any substance that decreases the surface tension of a liquid.

syndrome   The presence of several concurrent genetically engendered phenotypic abnormalities (e.g., Down syndrome is characterized by several observable defects, including mental retardation, the absence of a nasal bone, heart defects, a characteristic slanting of the eye, and often the closure of the intestine) in one person.

T

template   In biochemistry, a molecule or surface upon which another molecule is synthesized in complementary fashion, as in the synthesis of mRNA or the replication of DNA. The nucleotide pairs in DNA will separate forming two strands, one a template for the single-stranded mRNA molecule or both templates for the double-stranded DNA.

teratogens   The toxic or harmful agents (e.g. chemicals, viruses, radiation, or high fevers) responsible for disruptions during development, and hence birth defects.

teratology   The study of how environmental agents (e.g. chemicals, viruses, radiation, or high fevers) disrupt normal development.

testosterone   Hormone secreted by the Leydig cells in male embryos (XY embryos) that stimulates differentiation of the Wolffian duct and other secondary male characteristics. Development of the prostate gland and penis, and the changes seen during puberty are associated with testosterone.

therapeutic cloning   The use of somatic cell nuclear transplantation to create stem cells that are genetically identical to the person who will receive their cellular progeny. The goal of therapeutic cloning is to get stem cells that are genetically identical to the person needing treatment.

thymine (T)   One of the nitrogen-containing bases that make up DNA. See also **base; pyrimidine**.

totipotent   The condition of possessing all the genetic information and other capacities necessary to form an entire individual. Totipotent cells are usually early stage blastomeres that can produce any cell type of the conceptus, including trophoblast cells. Contrast with **pluripotent**.

totipotent stem cells   Stem cells that are capable of becoming any type of cell in the conceptus. Each cell of an early embryo, the first eight cells, is called a totipotent stem cell. Contrast with **pluripotent stem cells**.

transcription   The synthesis of RNA, using one strand of DNA as the template. The resulting single-stranded mRNA carries the code for a specific protein or set of proteins.

transcription factors   Proteins that are activated in the cytoplasm (i.e. a paracrine factor signaling the addition of a phosphate group) and that can then travel into the cell nucleus where they bind to the certain genes (in the area of the gene known as the enhancer), thus turning that gene on (or off).

transgene   A gene that has been inserted into the genome of another organism. The use of transgenes is one way to produce protein pharmaceuticals by inserting the human genes that encode for the desired protein into the zygote DNA of sheep, goats, or cows.

transgenic animals   An animal that has an additional gene incorporated into its genome. Such animals are used to produce protein pharmaceuticals by expressing the gene for a desired protein. A female that expresses the protein in her mammary tissue will secrete the protein in her milk.

translation   The synthesis of a protein (polypeptide). Translation takes place on ribosomes, using the information encoded in messenger RNA.

trimesters   The three divisions of the 9-month human gestation period. The first trimester is the time of organ development, the second and third involve growth and maturation of the fetus.

trophoblast   The outer layer of cells that attaches the embryo to the uterine wall, and develops into the fetal portion of the placenta. It does not form any of the tissues of the newborn. It is also called **trophectoderm** and **chorion**.

tubal pregnancy   A condition whereby the blastocyst implants into the oviduct wall. Also called an ectopic pregnancy

U

uracil (U)   One of the nitrogen-containing bases that make up RNA. See also **base; pyrimidine**.

uterus   Specialized portion of the female reproductive tract, that receives the embryo and nurtures embryo development. Also called the womb.

V

vas deferens   The tubular tract in males that carries sperm from the testes to the outside the body.

ventral   Toward or pertaining to the belly or lower side of an animal. Compare with **dorsal**.

W

Wolffian duct   Duct system seen in early embryos that can become the male epididymis and vas deferens (the tubular tract that carries sperm to the outside the body) in mammalian males.

womb   See uterus.

X

X-linked gene   A gene found only on the X chromosome; a sex-linked trait. A female has two X chromosomes, and thus may be a carrier for a genetic trait that is not observed in her phenotype; a male has one X chromosome, and if a gene on that chromosome is mutant, he will show the mutant phenotype.

Y

yolk sac   The extraembryonic (i.e., outside the embryo) membrane that forms from the inner cell mass in the early embryonic development of mammals, The embryo's early blood cells are made here. (Unlike bird yolk sacs, the mammalian yolk sac in fact contains no yolk).

Z

zona pellucida   The protein coat surrounding the egg. The sperm must get through this barrier to reach the egg, and the zona pellucida proteins bind to the sperm and induce the acrosome to release its enzymes. When the conceptus travels to the uterus, the zona pellucida prevents adhesion to oviduct walls (preventing an ectopic, or tubal, pregnancy). Upon entering the uterus, the embryo hatches from it.

zygote   The fertilized egg produced by the union of two gametes (the unfertilized egg from the female and the sperm from the male).

Quotation Sources and Literature Cited

Unit 1
Coleridge, S. T. 1885. *Miscellanea*. Bohn, London. p. 301.

Unit 2
Whitman, W. 1855. "Song of Myself." in *Leaves of Grass and Selected Prose*, S. Bradley (ed.), 1949, Holt Rinehart & Winston, New York, p. 25.

Unit 3
Schopenhauer, A. Quoted in Charles Darwin, 1871, *The Descent of Man*. Murray, London, p. 893.

Unit 4
Dawkins, R. 1997. "Thinking Clearly About Clones: How dogma and ignorance get in the way," *Free Inquiry*, Summer 1997. Vol, 17, No. 3.

Unit 5
Holub, M. 1990. "From the Intimate Life of Nude Mice." In *The Dimension of the Present Moment*, translated by D. Habova and D. Young. Faber & Faber, London, p. 38

Unit 6
New York Times. 1982. "Whether to Make Perfect Humans." Editorial, July 22, 1982.

Unit 7
Goethe, J. W. von. 1808. *Faust*, Part I. Lines 1–6.

Preface
Huxley, T. H. 1870. On Descartes' "Discourse Touching the Method of Using One's Reason Rightly and of Seeking Scientific Truth." *Macmillan's Magazine*, 24 March 1870; reprinted in T. H. Huxley, 1877, *Lay Sermons. Addresses, and Reviews*. Appleton, New York. p. 322.

Gibson, W. 1999. Quoted in M. Payser, "The Home of the Gay," *Newsweek,* March 1, 1999, p. 50.

Chapter 1
Aristotle. ca. 350 B.C.E. *Metaphysics*. Book 1, Part 2. D. Ross (trans.). Oxford University Press, New York, 1979.

Doyle, A. C. 1891. A Case of identity. *The Adventures of Sherlock Holmes*. Reprinted in *The Complete Sherlock Holmes Treasury* (1976). Crown Press, New York.

van Heyningen, V. 2000. Gene games of the future. *Nature* 408: 769–771.

Wolpert, L. 1986. *From Egg to Embryo: Determinative Events in Early Development*. Cambridge University Press, Cambridge, p. 41.

Chapter 2
Anderson, R. E. 2004. Ethics of embryonic stem cells: Letters to the editor. *New England Journal of Medicine*. 351: 1688.

Bonner, G. 1985. Abortion and early Christian thought. In J. H. Channer (ed.), *Abortion and the Sanctity of Human Life*. Paternoster Press, Exeter, UK.

Buss, M. 1967. The beginning of human life as an ethical problem. *Journal of Religion* 47: 244–255.

Cangiamila, F. E. 1758. *Embryologia Sacra*. Valenza, Palmero.

Donum vitae (Respect for Human Life). 1987. Instruction on respect for human life in its origin. Congregation for the Doctrine of the Faith (Vatican).

DeMarco, D. 1984. The Roman Catholic Church and abortion: A historical perspective, *Homoiletic Press & Pastoral Review* July 1984, pp. 59–66.

Dobzhansky, T. 1976. Living with the biological revolution. In R. H. Haynes (ed.), *Man and the Biological Revolution*, pp. 21–45. York University Press, Toronto.

Edelstein, L. 1967. *Ancient Medicine*. Johns Hopkins University Press, Baltimore.

Ford, N. M. 1988. *When Did I Begin? Conception of the Human Individual in History*. Cambridge University Press, New York.

Franklin, S. 1991. Fetal fascinations: New dimensions to the medical-scientific construction of fetal person-hood. In *Off–Center: Feminism and Cultural Studies*. (S. Franklin, C. Lury, and J. Stacey, eds.). Routledge, NY.

Gilbert, S. F. and R. Howes-Mischel. 2004. "Show me your original face before you were born": The convergence of public fetuses and sacred DNA. *History and Philosophy of Life Sciences* 26 (in press).

Grobstein, C. 1988. *Science and the Unborn: Choosing Human Futures*. Basic Books, New York.

Jakobovits, I. 1973. Jewish views on abortion. In D. Walbert and J. Butler (eds.), *Abortion, Society, and Law*. The Press of Case Western Reserve University, Cleveland and London.

Mange, E. J. and A. P. Mange. 1999. *Basic Human Genetics*. Sinauer Associates, Sunderland, MA.

McCormick, R. 1991. Who or what is a pre-embryo? *Kennedy Institute Bioethics Journal*. 1: 1–15.

Morowitz, H. J. and Trefil, J. S. 1992. *The Facts of Life: Science and the Abortion Controversy*. Oxford University Press, New York.

Musallam, B. 1990. The human embryo in Arabic scientific and religious thought. In G. R. Dunstan (ed.), *The Human Embryo: Aristotle and the Arabic and European Traditions*. University of Exeter Press, Exeter, UK.

O'Donovan, O. 1975. *The Christian and the Unborn Child*. Grove Books, Bramcote.

Petetchsky, R. 1987. Fetal images: The power of visual culture in the politics of reproduction. *Feminist Studies* 13: 263– 292.

Pinto–Correia, C. 1997. *The Ovary of Eve: Egg and Sperm and Preformation*. University of Chicago Press.

Ramsey P. 1970 Reference points in deciding about abortion. In J. T. Noonan (ed.), *The Morality of Abortion: Legal and Historical Perspectives*. Cambridge University Press, Cambridge.

Renfree, M. B. 1982. Implantation and placentation. In C. R. Austin and R. V. Short (eds.), *Reproduction in Mammals. 2. Embryonic and Fetal Development*, 2nd Ed. Cambridge University Press, Cambridge.

Rogerson, J.W. 1985. Using the Bible in the debate about abortion. In J. H. Channer (ed.), *Abortion and the Sanctity of Human Life*. Paternoster Press, Exeter, UK.

Sandel, M. J. 2004. Ethics of embryonic stem cells: Letters to the editor. *New England Journal of Medicine*. 351: 1689–1690.

Social Statements of the Evangelical Lutheran Church in America. www.elca.org/socialstatements/abortion/

Stabile, C. 1993. Shooting the mother. In P. A. Treichler, L. Cartwright and C. Penley (eds.), *The Visible Woman: Imaging Technologies, Gender, and Science*. New York University Press, New York.

Taylor, J. 1992. The public fetus and the family car: From abortion politics to a Volvo advertisement. *Popular Culture* 4 (2): 67–80.

Tertullian, *Apologia* (197 ce), Chapter 9. Translated by Rev. S. Thelwell, available at http://www.early christianwritings.com/text/tertullian01.html

Tooley, M. 1973. A defense of abortion and infanticide. In J. Feinberg (ed.), *The Problem of Abortion*. Wadsworth, Belmont, CA.

Tribe, L. 1990. *Abortion: The Clash of the Absolutes*. W.W. Norton & Company, New York.

Warren, M.A, 1999. The moral significance of birth. In J. A. Boss (ed.), *Analyzing Moral Issues*. McGraw–Hill, Boston.

Chapter 3

Arie, S. 2005. Cardinal says condoms could help stop AIDS. *The Guardian,* February 1, 2005. http://www. guardian.co.uk/pope/story/0,12272,1403083,00.html

Bradshaw, S. 2003. Vatican: Condoms don't stop AIDS. *The Guardian*, October 9, 2003. http://www. guardian.co.uk/aids/story/0,7369,1059068,00.html

Drazen, J., M. F. Greene and A. J. J. Woods. 2004. The FDA, politics, and Plan B. *New England Journal of Medicine* 350: 1561–1562.

Dunston, D. B., B. Columbo, and D. D. Baird. 2002. Changes with age in the leel and duration of fertility in the menstrual cycle. *Human Reproduction* 17: 1399–1403.

Grimes, D. A. 1997. Emergency contraception: Expanding opportunities for primary prevention. *New England Journal of Medicine* 337: 1077–1079.

Henshaw, S. K. 1998. Unintended pregnancy in the United States. *Family Planning Perspectives* 30: 24–29.

Jones, C. 2004. Druggists refuse to give out pill. *USA Today,* November 9, 2004.

Lillie, F. R. 1919. *Problems of Fertilization*. University of Chicago Press, Chicago.

MacDonald, G. J. 2004. Bishop Kevin Dowling: South African bishop battles church in battling AIDS. *National Catholic Reporter,* April 16, 2004.

Rivera, R., I. Yacobson and D. Grimes. 1999. The mechanism of action of hormonal contraceptives and intrauterine contraceptive devices. *American Journal of Obstetrics and Gynecology* 181: 1263–1269.

Speroff, L. and M. A. Fritz. 2005. *Clinical Gynecologic Endocrinology and Infertility*, 7th Ed. Lippincott Williams & Wilkins, Philadelphia.

Westhoff, C. 2003. Emergency contraception. *New England Journal of Medicine* 349: 1830–1839.

Wilcox, A., C. Weinberg and D. Baird. 1995. Timing of sexual intercourse in relation to ovultion: Effects on probability of conception, survival of the pregnancy, and sex of the baby. *New England Journal of Medicine* 333: 1517–1521.

Chapter 4

American College of Obstetricians and Gynecologists. 2004. Multiple gestation: Complicated twin, triplet, and high–order multifetal pregnancy. *ACOG Practice Bulletin* 104(4), October.

Andrews, L. B. 1999. *The Clone Age: Adventures in the New World of Reproductive Technology.* Henry Holt, New York.

Angier, N. 1999. Baby in a box. *New York Times.* May 16, 1999. Accessible at http://www.dpo.uab.edu/~svan/babybox.html

Beaujean, N. and 7 others. Non–conservation of mammalian preimplantation methylation dynamics. *Current Biology* 14: R266–R267

Bjerklie, D. 2005. Fit to be a mom? *Time* 165, No. 5 (January 31, 2005), p. 72.

Braude, P and P. Powell. 2003. Assisted conception. III: Problems with assisted conception. *British Medical Journal.* 327: 920–\923.

Callahan, T. L., J. E. Hall, S. L. Ettner, C. L. Christiansen, M. F. Greene and W. F. Crowley Jr. 1994. The economic impact of multiple–gestation pregnancies and the contribution of assisted–reproduction techniques to their incidence. *New England Journal of Medicine* 331: 244–249.

Caplan, A. 2005. Fertility clinics vary widely on who gets treatment. CNN Health. http://www.cnn.com/2005/HEALTH/01/19/fertility.ethics.ap/

CDC (Centers for Disease Control and Prevention, ASRM, SART). 2000. 2000 ART success rates. CDC, Atlanta, 2002. Quoted in Speroff and Fritz, 2005, *Clinical Gynecologic Endocrinology and Infertility*, 7th Ed. p. 1249.

Dawkins, R. 1976. *The Selfish Gene.* Oxford University Press, Oxford.

Donum vitae (Respect for Human Life). 1987. Instruction on respect for human life in its origin. Congregation for the Doctrine of the Faith (Vatican).

El–Shawarby, S. A., R. A. Margara, G. H. Trew, M. A. Laffan and S. A. Lavery. 2004. Thrombocythemia and hemoperitoneum after transvaginal oocyte retrieval for in vitro fertilization. *Fertil Steril.* 82: 735–737.

Ettner, S. L., C. L. Christiansen, T. L. Callahan and J. E. Hall. 1997. How low birthweight and gestational age contribute to increased inpatient costs for multiple births. *Inquiry* 34:325–339.

Gurmankin, A. D., A. L. Caplan and A. M. Braverman. 2005. Screening practices and beliefs of assisted reproductive technology programs. *Fertility and Sterility* 83: 61–67.

Hansen, M., J. J. Kurinczuk, C. Bower and S. Webb. 2002. The risk of major birth defects after intracytoplasmic sperm injection and *in vitro* fertilization. *New England Journal of Medicine* 346: 725–730.

Inhorn, M. 2003. quoted in www.eurekalert.org/pub_releases/2003–07/esfh–meb062503.php

ivf–infertility.com. 2004. Risks and complications of IVF treatment: Ovarian hyperstimulation syndrome. http://www.ivf–infertility.com/ivf/standard/complications/ovarian_stimulation/ohss.php

Jain, T., B. L. Harlow and M. D. Hornstein. 2002. Insurance coverage and outcomes of in vitro fertilization. *New England Journal of Medicine.* 347:661–666.

Keel, B. A. 1999. Quoted in L. C. Andrews, *The Clone Age: Adventures in the New World of Reproductive Technology.* Henry Holt, New York , p. 208–209.

Kwong, W. Y., A. E. Wild, P. Roberts, A. C. Willis, and T. P. Fleming. 2000. Maternal undernutrition during the preimplantation period of rat development causes blastocyst abnormalities and programming of postnatal hypertension. *Development* 127: 4195–4202.

Lewis, S. and H. Klonoff–Cohen. 2005. What factors affect intracytoplasmic sperm injection outcomes? *Obstet. Gyn. Survey* 60: 111–123.

McEvoy, T. G., K. D. Sinclair, L. E. Young, I. Wilmut and J. J. Robinson. 2000. Large offspring syndrome and other consequences of ruminant embryo culture in vitro: Relevance to blastocyst culture in human ART. *Human Fertlization* 3: 238–246.

Mitchell, A. A. 2002. Infertility treatment: More risks and challenges. *New England Journal of Medicine.* 346: 769–770.

Neumann, P. J., S. D. Gharib and M. C. Weinstein. 1994. The cost of a successful delivery with in vitro fertilization. *New England Journal of Medicine.* 331: 239–243.

Paul VI, Pope.; 1968. Encyclical letter *Humanae vitae*, No. 14, AAS 60 (1968), 488–489.

Robertson, J. 2004. Procreative liberty and harm to offspring in assisted reproduction. *American Journal of Law and Medicine* 30: 7–40.

Shea, J. B. 2004. The moral status of in vitro fertilization (IVF): biology and method. *Catholic Insight.* http://catholicinsight.com/online/church/vatican/article_475.shtml

Spar, D. 2005. Reproductive tourism and the regulatory map. *New England Journal of Medicine.* 352: 531–533.

Speroff, L. and M. A. Fritz. 2005. *Clinical Gynecologic Endocrinology and Infertility*, 7th Ed. Lippincott Williams & Wilkins, Philadelphia.

van Kooij, R. J., M. F. Peeters and E. R. te Velde. 1997. Twins of mixed races: consequences for Dutch IVF laboratories. *Hum. Reprod.* 12:2585–2587.

White, A. D. 1898. *A History of the Warfare of Science with Theology in Christendom.* NY: Appleton. Also at http://cscs.umich.edu/~crshalizi/White/air/rod.html

Wilson, E. O. 1975. *Sociobiology: The New Synthesis.* Harvard University Press, Cambridge, MA.

Winston, R. M. L. and K. Hardy. 2002. Are we ignoring potential dangers on in vitro fertilization and related treatments? *Nature Cell Biol.* 4: S14–S18.

Woodward, J. 1986. The non–identity problem. *Ethics* 96: 804–831.

Chapter 5

Ramachandran, R. 1999. In India, sex selection gets easier. *UNESCO Courier.* http://www.unesco.org/courier/1999_09/uk/dossier/txt06.htm

Weaver, T. 1999. New accuracy in sex selection. *Agricultural Research*, May 1999. http://www.ars.usda.gov/is/AR/archive/may99/accu0599.htm

Chapter 6

Al–Serour, G. A. 2000. *Ethical Implications of Human Embryo Research*. Islamic Educational, Scientific and Cultural Organization. http://www.isesco.org.ma /pub/Eng/Human%20Embryo/page10.htm.

Cardarelli, L. 1996. The Lost Girls. *Utne Reader*. 75:13–15.

Carmichael, M. 2004. No Girls, Please. *Newsweek*, January 26, 2004.

Center for Genetics and Society. 2002. Letter to American Society for Reproductive Medicine Regarding Sex Selection. http://www.genetics–and–society. org/resources/cgs/2002_asrm_sex_selection.html#2

Dahl, E. 2003. Should parents be allowed to use preimplantation genetic diagnosis to choose the sexual orientation of their children? *Human Reproduction* 18: 1368–1369.

Dahl, E., M. Beutel, B. Brosig and K. D. Hinsch. 2003. Preconception sex selection for non–medical reasons: A representative survey from Germany. *Human Reproduction* 18: 2231–2234.

Ethics Committee of the American Society for Reproductive Medicine Report: Preconception gender selection for nonmedical reasons. 2001. *Fertility and Sterility* 75: 861–864.

Ethics Committee of the American Society for Reproductive Medicine Report: Preimplantation genetic diagnosis and sex selection. 1999. *Fertility and Sterility*. 72: 595–598.

Giladi, A. 1990. Some Observations on the infanticide in Medieval Muslim Society. *International Journal of Middle East Studies*. 22(2): 185–200.

Gottlieb, S. 2001. US doctors say sex selection acceptable for non–medical reasons. *British Medical Journal*. 323(7317): 828.

Hudson, V.M. and A. Den Boer. 2004. *Bare Branches: The Security Implications of Asia's Surplus Male Population*. MIT Press, Cambridge, MA.

Human Genetics Alert Campaign Briefing. 2002. The case against sex selection. www.genetics–and–society.org/resources/sexselection/200212_hga.pdf

Islamic Organization for Medical Sciences Publication Series. 1983. The Full Minutes of the Seminar on Human Reproduction in Islam. May 24. Translated by M. Muneer S. Asbahi. www.islamset.com/ bioethics/firstvol.html

Jain, T., S. A. Missmer, R. S. Gupta and M. D. Hornstein. 2005. Preimplantation sex selection: Demand and preferences in an infertility population. *Fertility and Sterility* 83: 649–658.

Kalb, C. 2004. Brave new babies. *Newsweek*. January 26.

Leo, J. 1989. Baby boys, to order. *U.S. News & World Report*. 106(1): 59.

Li, L. M. 1991. Life and death in a Chinese famine: Infanticide as a demographic consequence of the 1935 Yellow River flood *Comparative Studies in Society and History*. July.

Ling, J. R. Preimplantation Genetic Diagnosis. http://users.aber.ac.uk/jrl/pgd.htm (University of Wales, Aberystwyth).

Liu, P. and G. A. Rose. 1996 Sex selection: The right way forward. *Human Reproduction* 1: 2343–2345.

Macklin, R. 1995. The ethics of sex selection. *Indian Journal of Medical Ethics*. 3: 61–64. http://www.issues inmedicalethics.org/034ed061.html

Mange, E. J. and A. P. Mange. 1999. *Basic Human Genetics*. Sinauer Associates, Sunderland, MA.

McDougall, J., D. DeWit and G. E. Ebanks. 1999. Parental preferences for sex of children in Canada. *Sex Roles* 41: 615–625.

Mudur, G. 2002. India plans new legislation to prevent sex selection. *British Medical Journal*. 324: 385.

Paul, D. B. 1995. *Controlling Human Heredity: 1865 to the Present*. Humanity Books/Prometheus Books, Amherst, NY.

Pebley, A. R. and C. F. Westoff. 1982. Women's sex preferences in the United States, 1970 to 1975. *Demography* 19: 177–189.

Pong, S.–L. 1994. Sex preference and fertility in Peninsular Malaysia. *Studies in Family Planning*. 25(3): 137–149.

The President's Council on Bioethics. 2003. Beyond Therapy: Controlling Sex of Children. Session 4. January 16. http://www.bioethics.gov/transcripts/ jan03/session4.html.

Ramachandran, R. 1999. In India, sex selection gets easier. *UNESCO Courier*. http://www.unesco.org/ courier/1999_09/uk/dossier/txt06.htm

Remaley, Rachel E. 2000. The Original Sexist Sin: Regulating Preconception Sex Selection Technology. *Case Western Reserve University Health Matrix: Journal of Law Medicine*. 10 Health Matrix 249.

Roberts, J.C. 2002. Customizing conception: A survey of preimplantation genetic diagnosis and the resulting social, ethical, and legal dilemmas. *Duke Law and Technology Review* www.law.duke.edu/journals /dltr/ articles/2002dltr0012.html.

Robertson, J. A. 2001. Preconception gender selection. *The American Journal of Bioethics*. 1(1): 2–5.

Robertson, J. A. 2003. Extending PGD: The ethical debate. *Human Reproduction* 18: 465–471.

Rorvik, D. and L. B. Shettles. 1970. *Your Baby's Sex: Now You Can Choose*, Dodd Mead & Company, Toronto.

Rosner, F. 1998. Judaism, genetic screening, and genetic therapy. *The Mount Sinai Journal of Medicine*. 65(5 and 6): 406–413.

Satpathy, R. and S. K. Mishra. 2000. The alarming "gender gap." *Bulletin of the World Health Organization*. 78(11): 1373.

Shalev, C. 2003. Weighing and Balancing the Values of a Jewish Democracy. Within and Beyond the Limits of Human Nature. Heinrich Boll Foundation, Berlin. October 12–15. www.gruene–akademie.de/down load/bioethik_shalev.pdf.

Shete, M.. 2005. Doc in the Dock. *Times of India* April 15, 2005.

Silver, L. M. 1998. A quandary that isn't: Picking a baby's sex won't lead to disaster. *Time*. 152(12): 83.

Smith, T.. 1993. Choosing sex. *British Medical Journal*. 307: 451.

Unnatural Selection. 1993. *The Economist,* January 30, 1993.

Vines, G. 1993. The hidden cost of sex selection. *New Scientist,* May 1, 1993.

Wadman, M. 2001. So You Want a Girl? *Fortune,* February 19, 2001.

Wahrman, M. Z. 2002. *Brave New Judaism: When Science and Scripture Collide.* Brandeis University Press, Waltham, MA.

Warren, M. A. 1985. *Gendercide: The Implications of Sex Selection* . Bowman & Allanhead Publishers.

Waters, B. 2003. Sex selection for social reasons: religious and moral perspectives. Sept. 24, 2003. http://www.eurekalert.org/pub_releases/2003–09/sari–ssf092303.php.

Wertz, D. C. and J. C. Fletcher. 1989. Fatal Knowledge? Prenatal diagnosis and sex selection. *The Hastings Center Report.* 19(3): 21–28.

Westoff, C. F. and R. R. Rindfuss. 1974. Sex preselection in the United States. *Science* 184: 633–636.

Winkvist, A. and H. Z. Akhtar. 2000. God should give daughters to rich families only: attitudes towards childbearing among low–income women in Punjab, Pakistan. *Social Science & Medicine.* 51(1): 73–82.

Chapter 7

Delage, Y. 1895. *La Structure du Protoplasma, les théories de Hérédité et les grandes problèms de la Biologie générale.* Scheicher, Paris. Translation quoted in J.–C. Beetschem and J.–L. Fischer, "Yves Delage (1854–1920) as a forerunner of modern nuclear transfer experiments." *International Journal of Developmental Biology* 48: 607–612 (2004).

Driesch, H. 1892. The potency of the first two cleavage cells in echinoderm development: Experimental production of partial and double formations. Reprinted in B. H. Willier and J. M. Oppenheimer (eds.), *Foundations of Experimental Embryology.* Hafner, New York, 1974.

Briggs, R. and T. J. King. 1952. Transplantation of living nuclei from blastula cells into enucleated frogs' eggs. *Proceedings of the National Academy of Sciences USA* 38: 455–464.

Hwang W.-S. and 14 others. 2004. Evidence of a pluripotent human embryonic stem cell line derived from a cloned blastocyst. *Science* 303: 1669–1674.

King, T. J. 1966. Nuclear transplantation in amphibia. *Methods in Cell Physiology* 2: 1–36.

King, T. J. and R. Briggs. 1956. Serial transplantation of embryonic nuclei. *Cold Spring Harbor Symposiums in Quantitative Biology* 21: 271–290.

McKinnell, R. G. 1978. *Cloning: Nuclear Transplantation in Amphibia.* University of Minnesota Press, Minneapolis.

Smith, L. D. 1956. Transplantation of the nuclei of primordial germ cells into enucleated eggs of *Rana pipiens. Proceedings of the National Academy of Sciences USA* 54: 101–107.

Spemann, H. 1918. Über die Determination der ersten Organanlagen des Amphibienembryo. *Wilhelm Roux Arch. Entwicklungsmech.* 43: 448–555.

Spemann, H. 1938. *Embryonic Development and Induction.* Yale University Press, New Haven.

Spemann, H. and O. Schotté. 1932. Über xenoplatische Transplantation als Mittel zur Analyse der embryonalen Induktion. *Naturwissenschaften* 20: 463–467.

Wilmut, I., K. Campbell and C. Tudge. 2000. *The Second Creation: Dolly and the Age of Biological Control.* Farrar, Straus, & Giroux, New York.

Chapter 8

AAAS Policy Brief: Human Cloning. 2003. American Association for Advancement of Science. http://www.aaas.org/spp/cstc/briefs/cloning/index.shtml.

AAMC (American Association of Medical Colleges). 2002. Letter in Support of "Human Cloning Prohibition Act" (S.1758). http://www.aamc.org/advocacy/library/research/corres/2002/012402.htm.

Andrews, L. B. 1999. *The Clone Age: Adventures in the New World of Reproductive Technology.* Henry Holt, New York.

Biever, C. 2003. UN postpones global human cloning ban. New Scientist Online News. http://www.newscientist.com/hottopics/cloning/cloning.jsp?id=ns99994359.

Blackburn, E. 2004. Bioethics and the political distortion of biomedical science. *New England Journal of Medicine* 350: 1379–1380.

Boiani, M., S. Eckardt, H. R. Schöler and K. J. McLaughlin. 2002. Distribution and level in mouse clones: consequences for pluripotency. *Genes and Development.* 16:1209–1219.

Brainard, J. and G. Blumenstyk. 2004. Two scientists on bioethics council say its reports favor Bush ideology. *Chronicles of Higher Education* (March 19) 50(28):A22.

Brainard , J. and S. Smallwood. 2004. Bush cuts two dissenters from federal bioethics advisory council. *Chronicles of Higher Education* (March 12) 50(27):A25.

Brock, D. W. 2002. Human cloning and our sense of self. *Science.* 296(5566):314–6.

Broyde, M. J. 2002. Cloning people and Jewish law: A preliminary analysis www.jlaw.com/Articles/cloning.html

Burley, J. 1999. The ethics of therapeutic and reproductive human cloning. *Cell and Developmental Biology* 10: 287–294.

Bush, G. W. 2001. Mission statement. Council of Bioethics. November 28, 2001, Federal Register date: November 30, 2001, Federal Register page: 66 FR 59851.

Butler, D. and M. Wadman. 1997. Calls for cloning ban sell science short. *Nature.* 386:8–9.

Campbell, C. S. 1997. Religious perspectives on human cloning in cloning human beings, National Bioethics Advisory Commission bioethics.georgetown.edu/nbac/pubs/cloning2/cc4.pdf

Caplan, A. 2001. Does ethics make a difference? In A. J. Klotzko (ed.), *The Cloning Sourcebook*. Oxford University Press, London.

Cook–Deegan, R. M. 1997. Do research moratoria work? In *Cloning Human Beings*, National Bioethics Advisory Commission. http://www.bioethics.georgetown.edu/nbac/pubs/cloning2/cc8.pdf

DiBerardino, M. A. and R. G. McKinnell. 1997. Backward compatible. *The Sciences*, Sept/Oct 1997: 32–37.

Garcia, J. A. 2000. Human cloning: Never and why not. In B. McKinnon (ed.), *Human Cloning: Science, Ethics and Public Policy*. University of Illinois Press, Chicago.

Gilbert, S. F. 1998. Human cloning. *New England Journal of Medicine* 339: 1558–1559.

Gillon, R. 2001. Human reproductive cloning. In A. J. Klotzko (ed.), *The Cloning Sourcebook*. Oxford University Press, London.

GPI (Genetics Policy Institute). 2004. http://www.genpol.org/index.htm

Grady, D. 2004. Debate over cloning in U.S. remains intense. *New York Times*, February 12, 2004.

Hall, S. S. 2004. Specter of cloning may prove a mirage. *New York Times* February 17, 2004.

Holden, C. 2004. Stem cells and slavery. *Science* 304: 1742.

Holm, S. 2001. A life in the shadow. In A. J. Klotzko (ed.), *The Cloning Sourcebook*. Oxford University Press, London.

Human Genetics Alert Campaign Briefing. 2002. The case against sex selection. www.genetics–and–society.org/resources/ sexselection/200212_hga.pdf.

Humpherys, D. and 7 others. 2001. Epigenetic instability in ES cells and cloned mice. *Science*. 293: 95–97.

Hwang W.-S. and 14 others. 2004. Evidence of a pluripotent human embryonic stem cell line derived from a cloned blastocyst. *Science* 303: 1669–1674.

Islamic View on Cloning. Recommendations of the Ninth Fiqh–Medical Seminar. Retrieved on March 22, 2004, from http://www.islamset.com/healnews/cloning/index.html.

Jaenisch, R. and I. Wilmut. 2001. Don't clone humans! *Science*. 291: 2552.

Janson-Smith, D. 2002. Human Reproductive Cloning Act. The Wellcome Trust. http://www.wellcome.ac.uk/en/genome/geneticsandsociety/hg15b009.html.

Kass, L. and J. Q. Wilson. 1998. *The Ethics of Human Cloning*. AEI Press, Washington, D.C.

Khosla, S., W. Dean, D. Brown, W. Reik and R. Feil. 2001. Culture of preimplantation mouse embryos affects fetal development and the expression of imprinted genes. *Biology of Reproduction*. 64: 918–926.

Kitchen, P. 2000. There will never be another you. In B. McKinnon (ed.), *Human Cloning: Science, Ethics and Public Policy*. University of Illinois Press, Chicago.

Kirkpatrick, R. 1999. Human cloning. In A. Weller (ed.), *The e–Journal of Human Cloning* http://www.humancloning.150m.com/article8.html.

Klotzko, A. J. (ed.). 2001. *The Cloning Sourcebook*. Oxford University Press, London.

Koerner, B. I. February 2002. Embryo police. *Wired Magazine*. 10.02.

Kolata, G. 2001. Researchers find big risk of defect in cloning animals. *New York Times*. 25 March 2001.

Lewontin, R. C. 2000. Cloning and the fallacy of biological determinism. In B. McKinnon (ed.), *Human Cloning: Science, Ethics and Public Policy*. University of Illinois Press, Chicago.

McLaren, A. 2000. Cloning: Pathways to a pluripotent future. *Science*. 288: 1775–1780.

McLaren, A. 2001. Dolly mice. In A. J. Klotzko (ed.), *The Cloning Sourcebook*. Oxford University Press, London.

McKinnell, R. G. and M. A. DiBerardino. 1999. The biology of cloning: History and rationale. *BioScience*. 49:875–885.

National Academy of Sciences (NAS). 2001a. U.S. Policy makers should ban human reproductive cloning. http://www4.nationalacademies.org/news.nsf/isbn/0309076374?OpenDocument

National Academy of Sciences (NAS). 2001b. Public funding of stem cell research enhances likelihood of attaining medical breakthroughs. http://www4.nationalacademies.org/news.nsf/isbn/0309076307?opendocument

National Academy of Sciences (NAS). 2002. Stem cells and the future of regenerative medicine. Committee on the Biological and Biomedical Applications of Stem Cell Research, Board on Life Sciences, National Research Council, Board on Neuroscience and Behavioral Health, Institute of Medicine. http://www.nap.edu/books/0309076307/html/

Nature. 2005. Missed opportunity to ban reproductive cloning. *Nature Cell Biology* 7: 323.

Nulman, I. and 8 others. 1997. Neurodevelopment of children exposed *in utero* to antidepressant drugs. *New England Journal of Medicine* 336: 258–262.

Ogonuki, N., K. Inoue, Y. Yamamoto, Y. Noguchi, K. Tanemura, O. Suzuki, H. Palaniswami and Acharya. 1997. Swami, Bill Clinton Has a Question: The President is asking all faiths, including Hinduism, about the ethics of human cloning. *Hinduism Today*. http://www.hinduismtoday.com/archives/1997/6/1997-6-05.shtml

Palaniswami, Acharya. 1997. Swami, Bill Clinton has a question: The President is asking all faiths, including Hinduism, about the ethics of human cloning. *Hinduism Today*. http://www.hinduismtoday.com/archives/1997/6/1997-6-05.shtml

Pew Research Center. 2002. Public makes distinctions on genetic research. http://people–press.org/reports/display.php3?PageID=408

Ratanakul, P. 2000. Buddhism, prenatal diagnosis, and cloning. Bioethics in Asia. (eds. N.Fujiki and D.R.J. Macer) Eubios Ethics Institute. p. 405–407.

http://www.biol.tsukuba.ac.jp/~macer/asiae/biae405.html.

Rideout, W. M., K. Eggan and R. Jaenisch. 2001. Nuclear cloning and epigenetic reprogramming of the genome. *Science* 293: 1093–1098.

Robertson, J. A. 1994. *Children of Choice: Freedom and the New Reproductive Technologies.* Princeton University Press, Princeton, NJ.

Robertson, J. A. 1998a. Human cloning and the challenge of regulation. *New England Journal of Medicine.* 339: 119–122.

Robertson, J. A. 1998b. Liberty, identity, and human cloning. *Texas Law Review.* 77: 1371–1456.

Sachedina, A. 2004. Islamic perspectives on cloning. *The e–Journal of Human Cloning.* http://www.human-cloning.150m.com/article9.html.

Seidel, G. Jr. 2000. Cloning mammals: Methods, applications and characteristics of cloned animals. In B. McKinnon (ed.), *Human Cloning: Science, Ethics and Public Policy.* University of Illinois Press, Chicago.

Simerly, C. and 11 others. 2003. Molecular correlates of primate nuclear transfer failures. *Science* 300: 297,n.

Singer, P. 2001. Cloning humans and cloning animals. In A. J. Klotzko (ed.), *The Cloning Sourcebook.* Oxford University Press, London.

Steinbock, B. 2000. Cloning human beings: Sorting through ethical issues. In B. McKinnon (ed.), *Human Cloning: Science, Ethics and Public Policy.* University of Illinois Press, Chicago.

Union of Concerned Scientists (UCS). 2004. Scientific integrity of policymaking. http://www.ucsusa.org/global_environment/rsi/page.cfm?pageID=1446

United States Conference of Catholic Bishops (USCCB), 1998. "Human Cloning Debate Raises Pro–Life Issues" www.usccb.org/prolife/issues/bioethic/fact598.htm (May 8, 2002)

United States Senate. 2001. To prohibit human cloning while preserving important areas of medical research, including stem cell research. 107th Congress, 1st Session. Senate Bill Number S. 1758. Washington.

United States Senate. 2002. To amend title 18, United States Code, to prohibit human cloning. 107th Congress, 2nd Session. Senate Bill Number S. 1899. Washington.

Wade, N. 2002. Moralist of science ponders its power. *New York Times,* March 19, 2002.

Wahrman, M. Z. 2002. *Brave New Judaism: When Science and Scripture Collide.* Braindeis University Press, London.

Wilson, F. John Stuart Mill. In *The Stanford Encyclopedia of Philosophy* (Fall 2003 Edition), Edward N. Zalta (ed.). http://plato.stanford.edu/archives/fall2003/entries/mill/

Zavos, P. 2001. Testimony before the House Subcommittee on Oversight and Investigation: Hearing on Issues Raised by Human Cloning Research. www.reproductivecloning.net/Articles/testimony.htm

Chapter 9

Assady, S., G. Maor, M. Amit, J. Itskovitz-Eldor, K. L. Skorecki and M. Tzukerman. 2001. Insulin production by human embryonic stem cells. *Diabetes* 50: 1691–1697.

Barberi, T. and 13 others. 2003. Neural subtype specification of fertilization and nuclear transfer embryonic stem cells and application in parkinsonian mice. *Nature Biotechnology* 21: 1200–1207.

Bjorklund, L. M. and 9 others. 2002. Embryonic stem cells develop into functional dopaminergic neurons after transplantation in a Parkinson rat model. *Proceedings of the National Academy of Sciences USA* 99: 2344–2349.

Brüstle, O. and 7 others. 1999. Embryonic stem cell–derived glial precursors: A source of myelinating transplants. *Science* 285: 754–756.

Butler, O. 1998. *The Parable of the Talents.* Warner Books, New York, p. 3.

Carvey, P.M., Z. D. Ling, C. E. Sortwell, M. R. Pitzer, S. O. McGuire, A, Storch and T. J. Collier. 2001. A clonal line of mesencephalic progenitor cells converted to dopamine neurons by hematopoietic cytokines: a source of cells for transplantation in Parkinson's disease. *Experimental Neurology* 171: 98–108.

Dor, Y., J. Brown, O. I. Martinez and D. A. Melton. 2004. Adult pancreatic beta–cells are formed by self–duplication rather than stem-cell differentiation. *Nature* 429: 1–6.

Edwards, R. G. 2004. Stem cells today. B1. Bone marrow stem cells. *Reproductive BioMedicine Online* 9 (5): 541–583.

Fraidenraich, D. and 6 others. 2004. Rescue of cardiac defects in *id* knockout embryos by injection of embryonic stem cells. *Science* 306: 247–252.

Hirata, Y. and 6 others. 2005. Human umbilical cord blood cells improve cardiac function after myocardial infarction. *Biochemical and Biophysical Research Communications* 327: 609–614.

Hwang W.-S. and 12 others. 2005. Patient–specific embryonic stem cells derived from human SCNT blastocysts. *Science* (May 19): DOI: 10.1126/science.1112286.

Jaenisch, R. 2004. Human cloning: The science and ethics of nuclear transplantation. *New England Journal of Medicine.* 351: 2787–1791.

Kaufman, D. S., E. T. Hanson, R. L. Lewis, R. Auerbach and J. A. Thomson. 2001. Hematopoietic colony–forming cells derived from human embryonic stem cells. *Proceedings of the National Academy of Sciences USA* 98: 10716–10721.

Khosrotehrani, K., K. L. Johnson, D. H. Cha, R. N. Salomon and D. W. Bianchi. 2004. Transfer of fetal cells with multilineage potential to maternal tissue. *Journal of the American Medical Association* 292: 75–80.

Kim, J. H. and 10 others. 2002. Dopamine neurons derived from embryonic stem cells function in an animal model of Parkinson's disease. *Nature* 418: 50–56.

Kogler, G. and 20 others. 2004. A new human somatic stem cell from placental cord blood with intrinsic pluripotent differentiation potential. *Journal of Experimental Medicine* 200: 123–135.

Korbling, M. and 7 others. 2002. Hepatocytes and epithelial cells of donor origin in recipients of peripheral–blood stem cells. *New England Journal of Medicine*. 346: 738–746.

LaBarge, M. A. and H. M. Blau. 2002. Biological progression from adult bone marrow to mononucleate muscle stem cell to multinucleate muscle fiber in response to injury. *Cell* 111: 589–601.

McDonald, J. W. and 7 others. 1999. Transplanted embryonic stem cells survive, differentiate, and promote recovery in injured rat spinal cord. *Nature Medicine* 5: 1410–1412.

National Institutes of Health. 2000. Stem cells: A primer. http://www.nih.gov/news/stemcells/primer.htm.

Osawa, M., K. Hanada, H. Hamada and H. Nakauchi. 1996. Long–term lymphohematopoietic reconstitution by a single CD34–low/negative hematopoietic stem cell. *Science* 273: 242–245.

Parsons, A. B. 2004. *The Proteus Effect: Stem Cells and their Promise for Medicine.* Joseph Henry Press, Washington, DC.

Pittenger, M. F. and 9 others. 1999. Multiulineage potential of adult human mesenchymal stem cells. *Science* 284: 143–147.

Rideout, W. M. III, K. Hochedlinger, M. Kyba, G. Q. Daley and R. Jaenisch. 2002. Correction of a genetic defect by nuclear transplantation and combined cell and gene therapy. *Cell* 109: 17–27.

Shamblott, M., J. Axelman, S. Wang, E. Bugg, J. Littlefield, P. Donovan, P. Blumenthal, G. Huggins and J. Gearhart, 1998. Derivation of pluripotent stem cells from cultured human primordial germ cells. *Proceedings of the National Academy of Sciences USA* 95: 13726–13731.

Silva, G. V. and 15 others. 2005. Mesenchymal stem cells differentiate into an endothelial phenotype, enhance vascular density, and improve heart function in a canine chronic ischemia model. *Circulation* 111: 150–156.

Slack, J. M. W. 1999. *Egg and Ego.* Springer Verlag, New York.

Solter, D., and J. Gearhart, 1999. Putting stem cells to work. *Science,* 283: 1468–1470.

Tae-gyu, K. 2004. Korean scientists succeed in stem cell therapy. *The Korean Times.* Nov. 26, 2004. Accessible at http://times.hankooki.com/lpage/200411/kt2004112617575710440.htm

Takagi, Y. and 19 others. 2005. Dopaminergic neurons generated from monkey embryonic stem cells function in a Parkinson primate model. *Journal of Clinical Investigation* 115: 102–109.

Wickelgren, I. 1999. Stem cells. Rat spinal cord function partially restored. *Science* 286: 1826–1827.

Chapter 10

American Association for the Advancement of Science. 2005. AAAS Policy Brief: Stem Cell Research. http://www.aaas.org/spp/cstc/briefs/stemcells/index.shtml

Berg, P. 2003. Drawing the line between ethical regenerative medicine research and immoral human reproductive cloning. Testimony U. S. Senate Judiciary Committee, March 19, 2003. Quoted in A. B. Parson, 2004. *The Proteus Effect: Stem Cells and their Promise for Medicine.* Joseph Henry Press, Washington, DC. p. 177.

BioScience. 2004. Patently problematic: the European patent laws may dissuade stem cell research. *BioScience 2004,* July 18–22. http://www.bioscience 2004.org/press/laurie.pdf.

Bush, G. W. 2001. Remarks on stem cell research August 9, 2001. http://www.whitehouse.gov/news/releases/2001/08/20010809–2.html

Callahan, D. 2003. *What Price Better Health?* University of California Press.

Chawengsaksophak, K., W. de Graaff, J. Rossant, J. Descamps and F. Beck. 2004. Cdx2 is essential for axial elongation in mouse development. *Proceedings of the National Academy of Sciences USA* 101: 7641–7645.

Commission of the European Communities, 2003. *Commission Staff Working Paper Report on Human Embryonic Stem Cell Research.* Brussels, 3,4,2003.

Cowan, C. A. and 10 others. 2004. Derivation of embryonic stem cell lines from human blastocysts. *New England Journal of Medicine* 350: 1353–1356.

Dhanda, R. 2002. *Guiding Icarus: Merging Bioethics with Corporate Interests.* Wiley–Liss, New York.

Dogin, J. L. 2004. Embryonic discourse: Abortion, stem cells, and cloning. *Issues in Law and Medicine* 19: 203–261.

Dorff, E. 1999. Testimony before NBAC. May 7, 1999. Washington, DC. Meeting transcript 18.

Feldt, G. 2004. *The War on Choice.* Bantam Books, New York.

Fiester, A., H. Scholer and A. Caplan. 2004. Stem cell therapies: It's time to talk to the animals. *Cloning Stem Cells* 6: 3–4.

Gearhart, J. 2004. Quoted in A. B. Parson, *The Proteus Effect: Stem Cells and their Promise for Medicine.* Joseph Henry Press, Washington, DC. p. 166.

Gilbert, S. 2003. ES cells, adult stem cells, and biotechnology. http://www.devbio.com/article.php?id =207.

Goldstein, L., 2000. Representing the American Society for Cell Biology. Stem Cell Testimony to the Labor, Health & Human Services and Education Subcommittee, 4/26/00. Retrieved July 2, 2004 from www.ascb.org/publicpolicy/goldsteinsctest.htm.

Hatch, O. 2001. Quoted in C. Connolly, Conservative pressure for stem cell funds builds. *Washington Post,* July 2, 2001, page A01.

Heschel, A. J. 1985. *The Insecurity of Freedom: Essays on Human Existence.* Schocken Books, New York.

Holland, S., 2001. Beyond the embryo: a feminist appraisal of the embryonic stem cell debate. In S. Holland, K. Lebacqz, and L. Zoloth (eds.). *The*

Human Embryonic Stem Cell Debate: Science, Ethics, and Public Policy. MIT Press, Cambridge, MA.

Hübner, K. and 9 others. 2003. Derivation of oocytes from mouse embryonic stem cells. *Sciencexpress*. www.sciencexpress.org / 1 May 2003 / Page 1/ 10.1126/science.1083452.

Hurlbut, W. B. 2004. Altered nuclear transfer as a morally acceptable means for the procurement of human embryonic stem cells. The President's Council on Bioethics, December 2004. http://www.bio ethics.gov/background/hurlbut.html.

Hwang W.-S. and 12 others. 2005. Patient–specific embryonic stem cells derived from human SCNT blastocysts. *Science* (May 19): DOI: 10.1126/ science.1112286.

Jha, V. 2004. Paid transplants in India: The grim reality. *Nephrology, Dialysis, Transplantation* 19: 541–543.

Jonas, H. 1969. Philosophical reflections on experimenting with human subjects. *Daedalus* 98: 219–247.

Kass, L. 2001. Why we should ban human cloning now: Preventing a Brave New World. New Republic, May, 2001.

Kennedy, D. 2005. Twilight for the Enlightenment? *Science* 308: 165.

Keown, D. 2004. *Buddhist Ethics: A Very Short Introduction*. Oxford University Press, Oxford.

Lanza, R., and N. Rosenthal. 2004. The stem cell challenge. *Scientific American* 290 (6): 92–99.

Magee, K. 2003. Ovarian hyperstimulation syndrome. *Healthwise*. California Pacific Medical Center, Boise, ID. http://www.cpmc.org/health/healthinfo/ index.cfm?section=healthinfo&page=article&sgml_i d=sto167456–sec.

Maienschein, J. 2003. *Whose View of Life Is It? Embryos, Cloning, and Stem Cells*. Harvard University Press, Cambridge, MA

Mann, C. 2005. The coming death shortage. *Atlantic Monthly*, May, pp. 92–102.

Marshall, E., 1998. Claim of human–cow embryo greeted with skepticism. *Science* 282: 1390–1391.

Melton, D. A., G. Q. Daley and C. G. Jennings. 2004. Altered nuclear transfer in stem-cell research: A flawed proposal. *New England Journal of Medicine* 351: 2791–2792.

Newman, S. A. 2002. The human chimera patent initiative. *Medical Ethics* 9: 2–7.

National Bioethics Advisory Commission, 1999. Ethical Issues in Human Stem Cell Research. Vol. 1. *Report and Recommendations of the National Bioethics Advisory Commission*. Rockville, MD.

National Institutes of Health, 2000. *National Institutes of Health Guidelines for Research Using Human Pluripotent Stem Cells*. http://stemcells.nih.gov/news/news Archives/stemcellguidelines.asp

Official Journal of the European Communities, 1998. Directive 98/44/Ec of the European Parliament and of the Council of 6 July 1998 on the Legal Protection of Biotechnological Inventions. 30.7.98, L 213/13. europa.eu.int/eur–lex/pri/en/ oj/dat/1998/l_213/ l_21319980730en00130021.pdf.

Phimister, E. G. and J. M. Drazen. 2004. Two fillips for human embryonic stem cells. *New England Journal of Medicine* 350: 1351–1352.

Pilcher, H., 2004. Britain's stem-cell store opens: Cell bank will serve researchers worldwide. *BioEd Online*. Retrieved August 16, 2004 from http://www.bio edonline.org/news/news.cfm?art=972.

President's Council on Bioethics. (2003) Staff working paper—Age retardation:sceintific possibilities and moral challenges. http://www.bioethics.gov/ background/age_retardation.html

Reagan, N. 2004. Speech to the Juvenile Diabetes Foundation, May 9. http://msnbc.msn.com/id/4937850/

Savulescu, J., 2000. The ethics of cloning and embryonic stem cells as a source of tissue for transplantation: time to take a positive approach to law reform in Australia. http://www.aph.gov.au/house/ committee/laca/humancloning/sub254.pdf.

Scheinfeld, R. and P. Bagley, 2001. The current state of embryonic stem cell patents. In Print. Retrieved August 12, 2004 from http://www.bakerbotts.com/news/printpage.asp? pubid=166231052001.

Shreeve, J. 2005. The other stem-cell debate. *New York Times* April 10, 2005. Available at http://www. genetics–and–society.org/newsdisp.asp?id=753.

Stolberg, C., 2001. Stem cell research advocates in limbo. *New York Times* January 20, 2001.

Timmons, H., 2004. Britain grants license to make human embryos for stem cells. *New York Times* August 12, 2004. http://www.nytimes.com/ 2004/08/12/science/12clone.html.

Wassarman, P. and G. Keller, (eds.), 2003. Differentiation of Embryonic Stem Cells. *Methods in Enzymology, Vol. 365*. Elsevier Academic Press, San Diego, CA.

Weiner, Y. 2005. Quoted in "The global race for stem cell therapies," BBC World, April 29. http://www.the-world.org/technology/stemcell/04.shtml

Whittaker, P., 2003. Stem cells, patents and ethics. Cardiff Centre for Ethics, Law and Society. http://www.ccels.cardiff.ac.uk/launch/ whittakerpaper.html.

Zoloth, L., 2001. Jordan's Banks: A view from the first years of human embryonic stem cell research. In S. Holland, K. Lebacqz, and L. Zoloth (eds.). *The Human Embryonic Stem Cell Debate: Science, Ethics, and Public Policy*. MIT Press, Cambridge, MA. Pp. 223–241.

Chapter 11

Blaese, R. M. and 9 others. 1995. T lymphocyte-directed gene therapy for ADA-SCID: Initial trial results after 4 years. *Science* 270: 475–480.

Chan, A. W. S., K. Y. Chong, C. Martinovich, C. Simerly and G. Schatten. 2001. Transgenic monkeys produced by retroviral gene transfer into mature oocytes. *Science* 291: 309–312.

Check, E. 2005. Trial of treatment for rare childhood illness is halted, again. *Nature* 433: 561.

Couzin, J. and J. Kaiser. 2005. Gene therapy. As Gelsinger case ends, gene therapy suffers another blow. *Science* 307: 1028.

Hartwell, L., L. Hood, M. L. Goldberg, A. E. Reynolds, L. M. Silver and R. C. Veres. 2004. *Genetics: From Genes to Genomes*, 2nd Ed. McGraw–Hill, New York.

Holtzman, N. A. and T. M. Marteau. 2000. Will genetics revolutionize medicine? *New England Journal of Medicine* 343: 141–144.

Huxley, J. 1957. Transhumanism. In *New Bottles for New Wine*. Chatto & Windus, London.

Lee, S. 1963. *The Amazing Spiderman*. Marvel Comics, New York.

Maimonides (Moshe ben Maimon). 1190. *A Guide for the Perplexed*. Translation by M. Friedlander, 1956. Dover, New York.

Pirottin, D., L. Grubet, A. Adamantidis, F. Farnir, C. Herens, H. Schröder and M. Georges. 2005. Transgenic engineering of male–specific muscular hypertrophy. *Proceedings of the National Academy of Sciences USA* 102: 6413–6418.

Roth, D. A., N. E. Tawa Jr., J. M. O'Brien, D. A. Treco and R. F. Selden. 2001. Nonviral transfer of the gene encoding coagulation factor VIII in patients with severe hemophilia A. *New England Journal of Medicine* 344: 1735–1742.

Schröder, A. R., P. Shinn, H. Chen, C. Berry, J. R. Ecker, F. Bushman. 2002. HIV-1 integration in the human genome favors active genes and local hotspots. *Cell* 110: 52–529.

Stolberg, S. G. 1999. The biotech death of Jesse Gelsinger. *New York Times Magazine* Nov. 28, 1999. p. 136–140, 149–150.

Chapter 12

AAAS. 2003. American Association for the Advancement of Science position paper. http://www.aaas.org/spp/dspp/sfrl/germline/report.pdf

Allen, G. E. 1996. Science misapplied: The eugenics age revisited. *Technology Review* 99 (Aug/Sep): 23–31.

Allen, G. E. 2001. Essays on science and society. Is a new eugenics afoot? *Science* 294: 59–61.

Begley, S. 1998. Designer babies: Altering unborn babies through gene therapy. *Newsweek*, November 9, 1998.

Belteki, G., M. Gertsenstein, D.W. Ow and A. Nagy. 2003. Site–specific cassette exchange and germline transmission with mouse ES cells expressing φC31 integrase. *Nature Biotechnology*. 21: 321–324

Bush, G. W. 2001. Mission statement, Council of Bioethics. November 28, 2001. Federal Register date: November 30, 2001, Federal Register page: 66 FR 59851.

Chapman, A. R. and M. S. Frankel. 2003. Framing the issues. In A. R. Chapman and M. S. Frankel (eds.), *Designing Our Descendents: The Promises and Perils of Genetic Modifications*. Johns Hopkins University Press, Baltimore. pp. 3–19.

Council for Responsible Genetics. 2001. Human Germline Manipulation Position Paper. www.gene-watch.org/educational/ germline _manipulation PP.pdf

Dobzhansky, T. 1962. *Mankind Evolving: The Evolution of the Human Species*. Yale University Press, New Haven, CT.

Dobzhansky, T. 1976. Living with the biological revolution. In R. H. Haynes (ed.), *Man and the Biological Revolution*, pp. 21–45. York University Press, Toronto.

Dowling, E. R. 1928. *Elementary Eugenics*. The University of Chicago Press, Chicago.

Duster, T. 2003. The hidden eugenic potential of germ–line interventions. In A. R. Chapman and M. S. Frankel (eds.), *Designing Our Descendents: The Promises and Perils of Genetic Modifications*. Johns Hopkins University Press, Baltimore. pp. 156–178.

Eugenics Archive. 2005. www.eugenicsarchive.org/eugenics/

Food and Drug Administration (FDA). 2000 "Tissue Action Plan: Tissue Reference Group Annual Report FY98. http://www.fda.gov/cber/tissue/tapfy98.htm

Food and Drug Administration (FDA). 2004. http://www.fda.gov/opacom/hpview.html

Gilbert, S. F. 2003. *Developmental Biology*. Sinauer Associates, Sunderland, MA.

Gonçalves, M. A. F. V. 2005. A concise peer into the background, initial thoughts, and practice of human gene therapy. *BioEssay* 27: 506–517.

Griffith, A. J. , W. Ji, M. E. Prince, R. A. Altschuler and M. H. Meisler. 1999. Optic, olfactory, and vestibular dysmorphogenesis in the homozygous mouse insertional mutant Tg9257. *Journal of Craniofacial Genetics and Developmental Biology* 19: 157–163.

Hayes, R. 2001. The quiet campaign for genetically engineered humans. *Earth Island Journal—News From Around the World*. V. 16, No. 1 (Spring 2001).

Irwin, A. 1997. A genetic breakthrough raises profound ethical problems as well as medical hopes. *The Daily Telegraph*, July 30 1997, p 14

Kevles, D. J. 1985. *In the Name of Eugenics*. Harvard University Press, Cambridge, MA.

Kolata, G. Scientists brace for changes in path of human evolution. *New York Times*, March 21 1998.

Leder, A., P. K. Pattengale, A. Kuo, T. A. Stewart and P. Leder. 1986. Consequences of widespread deregulation of the c–myc gene in transgenic mice: Multiple neoplasms and normal development. *Cell* 45: 485–495.

Ludmerer, K. M. 1972. *Genetics and American Society*. Johns Hopkins University Press, Baltimore.

Maienschein, J. 2003. *Whose View of Life Is It? Embryos, Cloning, and Stem Cells*. Harvard University Press, Cambridge, MA.

Mange, E. J. and A. P. Mange. 1999. *Basic Human Genetics*. Sinauer Associates, Sunderland, MA.

Mangles, J. Geneticists jump across ethical frontier. *National Review*. 21 May 2001

McPherron, A. C., A. M. Lawler and S. J. Lee. 1997. Regulation of skeletal muscle mass in mice by a new TGF–β superfamily member. *Nature* 387: 83.

Meek, J. IVF breakthrough modified genes. *Guardian Homes Pages*. 5 May 2001: pg 4.

Müller–Hill, B. 1988. *Murderous Science: Elimination by Scientific Selection of Jews, Gypsies, and Others, 1933–1945*. Oxford University Press, New York.

National Bioethics Advisory Commission, *Ethical Issues in Human Stem Cell Research*. Rockville, Maryland. September 1999: pg 93.

National Institutes of Health. 2003. http://www4.od. nih.gov/oba/rdna.htm

Newman, S. A. 2003. Averting the clone age: prospects and perils of human developmental manipulation. *Journal of Contemporary Health Law Policy*. 19: 431–463.

Paul, D.. 1996. *Controlling Human Heredity: 1865–Present*. Humanity Books/Prometheus Books, Amherst, NY.

Ponturo, D. 1998. Engineering the Human Germline: Summary Report. *Program on Science, Technology, and Society*. UCLA Center for the Study of Evolution and the Origin of Life. www.ess.ucla.edu/huge/report.html

Recombinant DNA Advisory Committee, "Guidelines for Research Involving Recombinant DNA Molecules," (Available from The Department of Health and Human Services, National Institutes of Health, http://www.nih.gov, Appendix M: NIH document entitled "Points to Consider in the Design and Submission of Protocols for the Transfer of Recombinant DNA into the Genome of Human Subjects," "Regulatory Issues: The Revised 'Points to Consider' Document" *Human Gene Therapy* 1 (1990), 93–103).

Rifkin, J. 1998. *The Biotech Century*. Putnam, New York.

Rockman, M. V., M. V. Hahn, N. Soranzo, D. Goldstein and G. A. Wray. 2003. Natural selection on human–specific transcription factor binding site regulating IL4 expression. *Current Biology* 130: 2118–2123 .

Schuelke, M. and 8 others. 2004. Myostatin mutation associated with gross muscle hypertrophy in a child. *New England Journal of Medicine* 350: 2682–2688.

Selden, S. 1999. *Inherited Shame: The Story of Eugenics and Racism in America*. Teachers College Press, Williston, VT.

Silver, L. 1998. *Remaking Eden: How Genetic Engineering and Cloning Will Transform the American Family*. Avon, New York.

Stock, G. 2002. *Redesigning Humans: Our Inevitable Genetic Future*. Houghton Mifflin, Boston.

Theological Letter Concerning the Moral Arguments of Biotechnology. 1983. Drafted by Jeremy Rifkin and signed by 58 leaders of the major religious denominations in the United States, June 8, 1983.

Tiemann–Boege, I., W. Navidi, R. Grewal, D. Cohn, D., B. Eskenazi, A. J. Wyrobek and N. Arnheim. 2002. The observed human sperm mutation frequency cannot explain the achondroplasia paternal age effect. *Proceedings of the National Academy of Sciences USA* 99: 14952–14957.

UNESCO. 2003. http://www.unesco.org/ibc/

Waters, B. 2002. European scholars support development of germ line modification. *EurekAlert*. http://www.eurekalert.org/pub_releases/2002–12/sari-ess121302.php

White, A. D. 1960. *A History of the Warfare of Science with Theology in Christendom*. (Reprint of the 1896 edition.) Dover, New York.

Chapter 13

Amundson, R. 2005. Disability, ideology, and quality of life: A bias to biomedical ethics. In D. Wasserman, R. Wachbroit and J. Bickenbach (eds.), *Quality of Life and Human Differences*. Cambridge University Press, New York. pp. 101–124.

Asch, A. 1999. Prenatal diagnosis and selective abortion: A challenge to practice and policy.*American Journal of Public Health* 89: 1649–1657.

Asch, A. 2003. Disability, equality, and prenatal testing: Contradictory or compatible? *Florida State Law Review* 30: 315–342.

Boorse, C. 1977. Health as a theoretical concept. *Philosophy of Science* 44: 542–573.

CDC (Center for Disease Control) 2005. U.S. Obesity Trends 1985–2003. http://www.cdc.gov/nccd-php/dnpa/obesity/trend/maps/

Colapinto, J. 2000. *As Nature Made Him: The Boy Who Was Raised As a Girl*. HarperCollins, New York.

Davis, L. 1995. *Enforcing Normalcy: Disability, Deafness, and the Body*. Verso Press, London.

Dennis, C. 2004. Deaf by design. *Nature* 431: 894896.

Dobzhansky, T. (1962). *Mankind Evolving*. Yale University Press, New Haven, CT.

Dreger, A. 1998. Ambiguous sex or ambivalent medicine? The Hastings Center Report 28: 24–35.

Dreger, A. 1999. When medicine goes too far in the pursuit of normality. *Health Ethics Today*. Provincial Health Ethics Network. 10(1). http://www.phen.ab.ca/materials/het/het10–01a.html

Dreger, A. 2004. Raising kids in a "third gender." Intersex Society of North America. http://www.isna.org/drupal/node/view/565

Fausto–Sterling, A. 2000. *Sexing the Body: Gender Politics and the Construction of Sexuality*. Basic Books, NY.

Finn, Robert J. 1998. *Sound from Silence: Development of Cochlear Implants*. National Academy of Sciences Press, Washington, DC.

Futuyma, D. 2005. *Evolution*. Sinauer Associates, Sunderland, MA.

Gilbert, S. F. 1979. The metaphorical structuring of social perceptions. *Soundings* 62: 166–186.

Gould, S. J. and R. Lewontin. 1994. The spandrels of San Marco and the panglossian paradigm: A critique of the adaptationist programme In E. Sober (ed.), *Conceptual Issues in Evolutionary Biology*. MIT Press, Cambridge, MA.

Intersex Society of North America. 2004. Frequency: How Common are Intersex Conditions? http://www.isna.org/drupal/node/view/91

Hopps, H. C. 1964. *Principles of Pathology*, 2nd Ed. Appleton–Century–Crofts, New York.

Keller, H. 1927. Dreams that come true. *Personality*, December 1927.

King, C. D. 1945. The meaning of normal. *Yale Journal of Biology and Medicine* 17: 493–501.

MacDonald, M. 2000. Intersex and gender identity. http://www.ukia.co.uk/voices/is_gi.htm

Muirhead, E. S. and S. James. 1999. Commentary: Normal vs. Abnormal. *Health Ethics Today*. Provincial Health Ethics Network. http://www.phen.ab.ca/materials/het/het10–01b.html

Murphy, E. A. 1972. The normal and the perils of the sylleptic argument. *Perspectives in Biology and Medicine* 15: 566–582.

Nordenfelt, L. 1985. A sketch for a theory of health. *Acta Philosophica Fennica* 38: 203–217.

Nordenfelt, L. 1995. *On the Nature of Health: An Action–Theoretic Approach*, 2nd Ed. Kluwer Academic Publishers, Dordrecht, Netherlands.

Reznek, L. 1987. *The Nature of Disease*. Routledge & Kegan Paul, New York.

Tieffer, L. 2004. Are drugs really the answer to women's sexual "problems"? *Genetic Engineering News* 24: 6–8.

United Kingdom Intersex Association (UKIA). Recommendations for the Assessment and Treatment of Intersex Conditions. http://www.ukia.co.uk/ukiaguid.htm

Vachá, J. 1978. Biology and the problem of normality. *Scientia* 72: 823–846.

Vachá, J. 1985. German constitutional doctrine in the 1920s and 1930s and pitfalls of the contemporary conception of normality in biology and medicine. *Journal of Medicine and Philosophy* 10: 339–367.

Watson, J. D. 1996. President's essay: Genes and politics. Annual Report of Cold Spring Harbor Laboratories, 1996. Quoted in Asch 1999, pp. 1651–1652.

Wertz, D. 2002. *Society and the Not-so-New Genetics: What are we Afraid of? Some Future Predictions from a Social Scientist*. The Shriver Center. http://www.umassmed.edu/shriver/Research/SocialScience/Staff/Wertz/lawjrl.htm

Chapter 14

Adler, F. R. and Harvell, C. D. 1990. Inducible defenses, phenotypic variability, and biotic environments. *Trends in Ecology & Evolution* 5: 407–410.

Agrawal A.A., Laforsch, C, and Tollrian, R. 1999. Transgenerational induction of defenses in animals and plants. *Nature* 401: 60–63.

Barker, D. J. P. 1994a. The fetal origins of adult disease. *Proceedings of the Royal Society of London*, Series B 262: 37–43.

Barker, D. J. P. 1994b. *Mothers, Babies, and Disease in Later Life*. BMJ Publishing, London.

Black, J. E., Issacs, K. R. Anderson, B. J. Alcantara, A. A. and Greenough, W. T. 1990. Learning causes synaptogenesis, whereas motor activity causes angiogenesis, in cerebellar cortex of adult rats. *Proceedings of the National Academy of Sciences USA* 87: 5568–5572.

Bogin, B. 1997. Evolutionary hypotheses for human childhood. *Yearbook of Physical Anthroppology* 40: 63–89.

Cebra, J. J. 1999. Influences of microbiota on intestinal immune system development. *American Journal of Clinical Nutrition* 69 (Suppl.): 1046S–1051S.

Childs, B. 1999. *Genetic Medicine*. Johns Hopkins University Press, Baltimore.

Dawkins, R. 1976. *The Selfish Gene*. Oxford University Press, Oxford.

Dawkins, R. 1986. *The Blind Watchmaker*. W. W. Norton, New York.

Emlen, D. J. 1997. Diet alters male horn allometry in the beetle *Onthophagus acuminatus* (Coleoptera: Scarabaeidae). *Proceedings of the Royal Society of London*, Series B 264: 567–574

Farber, S. 1981. *Identical Twins Reared Apart: A Reanalysis*. Basic Books, New York.

Frost, N. 2004. Quoted in A. B. Parson, *The Proteus Effect*. John Henry Press, DC. pp. 152–153.

Gage, F. H. et al., 1995. Survival and differentiation of adult neuronal progenitor cells transplanted into an adult brain. *Proceedings of the National Academy of Sciences USA* 92: 11879–11883.

George, R. 2002. Cloning addendum. http://www.nationalreview.com/document/document071602.asp

Gilbert, S. F. 2001. Ecological developmental biology: developmental biology meets the real world. *Developmental Biology* 233: 1–12.

Gilbert, S. F. 2003. *Developmental Biology*, 7th Ed. Sinauer Associates, Sunderland, MA.

Gilbert, W. 1990. A vision of the Grail. In D. J. Kevles and L. Hood (eds.), *The Code of Codes*. Harvard University Press, Cambridge, MA. p. 96.

Gluckman, P. and M. Hanson. 2005. *The Fetal Matrix: Evolution, Development, and Disease*. Cambridge University Press, New York.

Gould, S. J. 1977. *Ontogeny and Phylogeny*. Belknap Press, Cambridge, MA.

Gould, S. J. 1997. Individuality. *The Sciences* 37: 14–16.

Harris, R. F. 1997. Journal journalism. *Current Biology* 7: 458.

Hooper, L.V., L. Bry, P. G. Falk and J. I. Gordon. 1998. Host-microbial symbiosis in the mammalian intestine: exploring an internal ecosystem. *BioEssays* 20: 336–343.

Jablonka, E. and M. J. Lamb. 2002. The changing concept of epigenetics. *Annals of the New York Academy of Science* 981: 82–96.

Johannsen, W. 1903. *Über Erblichkeit in Populationen und in reinen Linien*. Gustav Fisher, Jena.

Keller, E. F. 1992. Nature, nurture, and the human genome project. In D. J. Kevles and L. Hood (eds.), *The Code of Codes*. Harvard University Press, Cambridge, MA. pp. 281–299.

Koshland, D. 1989. Editorial. *Science* 246: 189.

Lewontin, R. 2001. *The Triple Helix: Gene, Organism, and Environment*. Harvard University Press, Cambridge, MA.

Lindee, M. S. 2005. *Moments of Truth in Genetic Medicine*. Johns Hopkins University Press, Baltimore.

Mann, C. C. 1984. Behavioral genetics in transition. *Science* 264: 1686–1689.

Martin, R. D. 1990. *Primate Origins and Evolution: A Phylogenetic Reconstruction*. Princeton University Press, Princeton, NJ.

Montagu, A. 1989. *Growing Young*. Bergin and Garvey, Mass.

Nelkin, D. and M. S. Lindee. 1995. *The DNA Mystique: The Gene as a Cultural Icon*. W. H. Freeman, New York.

Nijhout, H. F. 1991. *The Development and Evolution of Butterfly Wing Patterns*. Smithsonian Institution Press, Washington, DC.

Norman, A. and P. Harper. 1989. Survey of manifesting carriers of Duchenne and Becker muscular dystrophy in Wales. *Clinical Genetics* 36: 31–37.

Paglia, C. 1992. *Sex, Art, and American Culture*. Vintage Press, New York. pp. 103–104.

Perry, D. and R. Good. 1968. Experimental arrest and induction of lymphoid development in lympho epithelial tissues of rabbits. *Laboratory Investigator* 18: 15–26.

Pieau, C., Girondot, N., G. Richard-Mercier, M. Desvages, P. Dorizzi and P. Zaborski. 1994. Temperature sensitivity of sexual differentiation of gonads in the European pond turtle. *Journal of Experimental Zoology* 270: 86–93.

Portmann, A. 1941. Die Tragzeiten der Primaten und die Dauer der Schwangerschaft beim Menschen: ein Problem der vergleichen Biologie. *Rev. Suisse Zool.* 48: 511–518.

Purves, D. and Lichtman, J. W. 1985. *Principles of Neural Development*. Sinauer Associates, Sunderland, MA.

Rawls, J. F., B. F. Samuel and J. I. Gordon. 2004. Gnotobiotic zebrafish reveal evolutionarily conserved responses to the gut microbiota. *Proceedings of the National Academy of Sciences USA* 101: 4596–4601.

Rhee, K.-J, P. Sethupathi, A. Driks, D. K. Lanning and K. L. and Knight. 2004. Role of commensal bacteria in development of gut–associated lymphoid tissues and pre-immune antibody repertoire. *Journal of Immunology* 172: 1118–11124.

Richards, C. S.and 8 others. 1990. Skewed X–inactivation in a female MZ twin results in Duchenne muscular dystrophy. *American Journal of Human Genetics* 46: 672–681.

Rook, G. A. and Stanford, J. L. 1998. Give us this day our daily germs. *Immunology Today* 19: 113–116.

Rose, S. 1997. *Lifelines: Biology, Freedom, Determinism*. Penguin Books, New York.

Stappenbeck, T. S., L. V. Hooper, and J. I. Gordon. 2002. Developmental regulation of intestinal angiogenesis by indigenous microbes via Paneth cells. *Proceedings of the National Academy of Sciences USA* 99: 15451–15455.

Steidler, L. 2001. Microbiological and immunological strategies for treatment of inflammatory bowel disease. *Microbes and Infection* 3: 1157–1166

Suhonen, J. O., Peterson, D. A., Ray, J., and Gage, F. H. 1996. Differentiation of adult hippocampus–derived progenitors into olfactory neurons in vivo. *Nature* 383: 624–627.

Taylor, A. 2004. The consequences of selective participation on behavioral genetic findings: Evidence from simulated and real data. *Twin Research* 7: 485–504.

Tollrian, R. and C. D. Harvell. 1999. *The Ecology and Evolution of Inducible Defenses*. Princeton University Press, Princeton, NJ.

Turner, A. M. and W. T. Greenough. 1983. Synapses per neuron and synaptic dimensions in occipital cortex of rats reared in complex, social, or isolation housing. *Acta Stereologica* 2 (Suppl. 1): 239–244.

Umesaki ,Y. 1984. Immunohistochemical and biochemical demonstration of the change in glycolipid composition of the intestinal epithelial cell surface in mice in relation to epithelial cell differentiation and bacterial association. *J. Histochem.Cytochem.* 32: 299–304.

Warner, R. R. 1993. Mating behavior and hermaphroditism in coral reef fishes. In P. W. Sherman and J. Alcock, (eds), *Exploring Animal Behavior*. Sinauer Associates, Sunderland, MA, pp. 188–196.

Waterland, R. A. and Jirtle, R. L. 2003a. Transposable elements: Targets for early nutritional effects of epigenetic gene regulation. *Molecular Cell Biology* 23: 5293–5300.

Waterland, R. A. and Jirtle, R. L. 2003b. Developmental relaxation of insulin–like growth factor 2 imprinting in the kidney is determined by weanling diet. *Pediatric Research* 53 (Suppl) 5A.

Waterland, R.A. and Jirtle, R.L. 2004. Early nutrition, epigenetic changes at transposons and imprinted genes, and enhanced susceptibility to adult chronic diseases. *Nutrition* 20: 63–68.

Watson, J. 1997. Quoted in B. Loudon, Geneticist attacked over "abort gays" view. *The Advertiser*, Feburary 17, 1997.

Weaver, I. C. and 8 others. 2004. Epigenetic programming by maternal behavior. *Nature Neuroscience* 7: 847–854.

Wickelgren, I. 1999. Nurture helps mold able minds. *Science* 283: 1832–1834.

Chapter 15

Amico, J. A., R. C. Mantella, R. R. Vollmer and X. Li . 2004. Anxiety and stress responses in female oxytocin deficient mice. *Journal of Neuroendocrinology* 16: 319–324.

Arluke, A. and C. R. Sanders. 1996. *Regarding Animals*. Temple University Press. Philadelphia.

Asquith, Pamela. 1983. The monkey memorial services of Japanese primatologists. *Royal Anthropological Institute News* 54: 3–4.

Bekoff, Marc. 2002a. Cognitive ethology, deep ethology, and the great ape/animal project: expanding the community of equals. In J. P. Gluck, T. DiPasquale and F. B. Orlans (eds.), *Applied Ethics in Animal Research: Philosophy, Regulation, and Laboratory Applications*. Purdue University Press, West Lafayette, IN.

Bekoff, Marc. 2002b. *Minding Animals: Awareness, Emotions, and Heart*. Oxford University Press. Oxford.

Bentham, J. 1789. *Principles of Morals and Legislation*. Reprinted by Hafner, New York (1948).

Biller-Andorno, Nikola. 2002. Can they reason? Can they talk? Can we do without moral price tags in animal ethics? In J. P. Gluck, T. DiPasquale and F. B. Orlans (eds.), *Applied Ethics in Animal Research: Philosophy, Regulation, and Laboratory Applications*. Purdue University Press, West Lafayette, IN.

Block, B. M., R. W. Hurley and S.N. Raja. 2004. Mechanism-based therapies for pain. *Drug News Perspectives* 17(3):172–186.

Bower, B. 2004. Teen brains on trial: The science of neural development tangles with the juvenile death penalty. *Science News*. 165(19): 299–301.

Brainard, G. 1988. Illumination of laboratory animal quarters: Participation of light irradiance and wavelength in the regulation of the neuroendocrine system. In H. N. Guttman, J. A. Mench, R. C. Simmonds (eds.), *Science and Animals: Addressing Contemporary Issues*. Conference of Scientists Center for Animal Welfare in Washington, DC. June 22–25.

Choi, H. 1998. Koreans honor dead lab animals. *Wall Street Journal*, November 10, 1998, p. B1.

Cohen, C. 2001. In defense of the use of animals. In C. Cohen and T. Regan (eds.), *The Animal Rights Debate*. Rowman & Littlefield, Lanham, MD.

Cohen, C. and T. Regan. 2001. *The Animal Rights Debate*. Rowman & Littlefield, Lanham, MD.

CIOMS (Council for International Organizations of Medical Sciences). 1985.International Guiding Principles for Biomedical Research involving Animals. Geneva, Switzerland. http://www.cioms.ch/1985_texts_of_guidelines.htm

Darwin, C. 1871. *The Descent of Man*. Ch. 3. Quoted in Darwin (ed. P. Appleman) Norton, New York, p. 245.

DeGrazia, D. 2002. *Animal Rights: A Very Short Introduction*. Oxford University Press Inc. New York.

Denver, R. J. 1997. Environmental stress as a developmental cue: Corticotropin-releasing hormone is a proximate mediator of adaptive phenotypic plasticity in amphibian metamorphosis. *Hormones and Behavior* 31: 169–179.

de Sousa, R. 1980. Arguments from nature. *Zygon* 15: 169–191.

Fiorillo, C. D., P. N. Tobler and W. Schultz. 2003. Discrete coding of reward probability and uncertainty by dopamine neurons. *Science*. 299: 1898–1902.

Foltz, R.C. 2004. Islam, animals, and vegetarianism. *Encyclopedia of Religion and Nature*. http://www.religionandnature.com/encyclopedia/samples/Islam_and_Animals.htm.

Fossey, D. 1983. *Gorillas in the Mist*. Houghton-Mifflin, Boston.

Fox, M. A. 1986. *The Case for Animal Experimentation: An Evolutionary and Ethical Perspective*. University of California Press, Berkeley.

Francione, G. B. 2000. *Introduction to Animal Rights: Your Child or the Dog?* Temple University Press. Philadelphia.

French, R. D. 1975. *Antivivisection and Medical Science in Victorian Society*. Princeton University Press. Princeton.

Frey, R. G. 2002. Ethics, Animals, and Scientific Inquiry. In J. P. Gluck, T. DiPasquale and F. B. Orlans (eds.). *Applied Ethics in Animal Research: Philosophy, Regulation, and Laboratory Applications*. Purdue University Press, West Lafayette, IN.

Garner, R. 1993. *Animals, Politics, and Morality*. Manchester University Press. Manchester.

Gewin, V. 2002. Welfare amendment omits lab animals. *Nature*. 417: 106.

Gleyzer, R., A. R. Felthous and C. E. Holzer 3rd. 2002. Animal cruelty and psychiatric disorders. *Journal of the American Academy of Psychiatry and Law* 30(2):257–265.

Gluck, J. P. and T. DiPasquale. 2002. Introduction. In J. P. Gluck, T. DiPasquale and F. B. Orlans (eds.). *Applied Ethics in Animal Research: Philosophy, Regulation, and Laboratory Applications*. Purdue University Press, West Lafayette, IN.

Goodall, J. 1988. *In the Shadow of Man*. Houghton-Mifflin, Boston.

Guttman, H. N., J. A. Mench and R. C. Simmonds (eds.). 1988. *Science and Animals: Addressing Contemporary Issues*. Conference of Scientists Center for Animal Welfare in Washington, DC. June 22–25. Question and Answer Session.

Haraway, Donna. 1989. *Primate Visions: Gender, Race, and Nature in the World of Modern Science*. Routledge. New York.

Haraway, D. J. 2003. *The Companion Species Manifesto: Dogs, People, and Significant Otherness*. Prickly Paradigm Press, Chicago. P. 50.

Hebebrand, J., W. Friedl, R. Reichelt, E. Schmitz, P. Moller, and P. Propping. 1988. The shark GABA benzodiazepine receptor: Further evidence for a not-so-late phylogenetic appearance of the benzodiazepine receptor. *Brain Research* 446: 251–261.

Hiebert, S. M., M. Ramenofsky, K. Salvante, J. C. Wingfield, and C. L. Gass. 200. Noninvasive methods for measuring and manipulating corticosterone in hummingbirds. *General and Comparative Endocrinology* 120: 235–247.

Holden, C. 2004. SARS war memorial. *Science* 306: 2184.

Iliff, Susan. 2002. An additional "R": Remembering the animals. *Institute for Laboratory Animal Research Journal*, January, pp. 38–47.

Inoue, K, E. P. Zorrilla, A. Tabarin, G. R. Valdez, S. Iwasaki, N. Kiriike and G. F. Koob. 2004. Reduction of anxiety after restricted feeding in the rat: implication for eating disorders. *Biological Psychiatry* 55: 1075–1081.

Institute of Laboratory Animal Resources Commission on Life Sciences. 1996. *Guide for the Care and Use of Laboratory Animals*. National Research Council. National Academy Press. Washington, DC. http://www.nap.edu/readingroom/books/labrats/

Kast, A. 1994. Memorial stones for the souls of animals killed for human welfare in Japan. *Berliner-Muenchener Tierarztliche Wochenschrift* May: 166–171.

Katahara-Frish, J. 1991. Culture and primatology: East and west. In L. Fedigan and P. Asquith (eds.), *The Monkeys of Arashiyama: Thirty-five Years of Research in Japan and the West*. State University of New York Press, Albany. pp. 74–80.

Keck, M. E., F. Holsboer and M. B. Muller. 2004. Mouse mutants for the study of corticotropin-releasing hormone receptor function: development of novel treatment strategies for mood disorders. *Annals of the New York Academy of Science* 1018: 445–457.

Kennaway, S. 1980. Classroom use of animals for scientific purposes. In *Buddhism and Respect for Animals: Collected Excerpts from the Journal of Shasta Abbey*. Mt. Shasta, CA. Vol. XI, nos. 7 and 8.

Kest, B., E. Sarton and A. Dahan. 2000. Gender differences in opioid-mediated analgesia: animal and human studies. *Anesthesiology*. 93: 539–547.

Levy, Z. 1996. Ethical issues of animal welfare in Jewish thought. *Judaism* 45: 47–57

Link, K. H., M. Kornmann, G. H. Leder, U. Butzer, J. Pillasch, L. Staib, F. Gansauge and H.G. Beger. 1996. Regional chemotherapy directed by individual chemosensitivity testing in vitro: A prospective decision-aiding trial. *Clinical Cancer Re*search 2: 1469–1474.

Linzey, A. 1994. *Animal Theology*. University of Illinois Press. Chicago.

Masri Al-Hafiz, B.A. 2004. *Animals in Islam*. http://www.chai-online.org/islam.htm.

Midgley, M. 1983. *Animals and Why They Matter*. University of Georgia Press, Athens.

Milius, S. 2003. Unfair trade: Monkeys demand equitable exchanges. *Science News* 164(12):181.

Molina-Hernandez, M. and N. P. Tellez-Alcantara. 2004. Rats socially reared and full fed learned an autoshaping task, showing less levels of fear-like behaviour than fasted or singlyreared rats. *Laboratory Animals* 38(3): 236–245.

Morton, D. 2002. Importance of non-statistical experimental design. In J. P. Gluck, T. DiPasquale and F. B. Orlans (eds.), *Applied Ethics in Animal Research: Philosophy, Regulation, and Laboratory Applications*. Purdue University Press, West Lafayette, IN.

Muscari, M. 2004. Juvenile animal abuse: Practice and policy implications for PNPs. *Journal of Pediatric Health Care* 18: 15–21.

Ness, R. B., S. R. Wisniewski, H. Eng and W. Christopherson. 2002. Cell viability assay for drug testing in ovarian cancer: In vitro kill versus clinical response. *Anticancer Research* 22(2B): 1145–1149.

Nielsen, M, C. Braestrup and R. F. Squires. 1978. Evidence for a late evolutionary appearance of brain-specific benzodiazepine receptors: An investigation of 18 vertebrate and 5 invertebrate species. *Brain Research* 141: 342–346.

Orlans, F. B. 2002. Ethical themes of national regulations governing animal experiments: An international perspective. In J. P. Gluck, T. DiPasquale and F. B. Orlans (eds.), *Applied Ethics in Animal Research: Philosophy, Regulation, and Laboratory Applications*. Purdue University Press, West Lafayette, IN.

Patterson, F. and E. Linden. 1985. *The Education of Koko*. Holt, Rinehart and Winston, New York.

Phillips, M. T. and J. A. Sechzer. 1989. *Animal Research and Ethical Conflict*. Springer-Verlag, New York.

Porter, D. 1988. Value versus animal cost. In H. N. Guttman, J. A. Mench, R. C. Simmonds (eds.), *Science and Animals: Addressing Contemporary Issues*. Conference of Scientists Center for Animal Welfare in Washington, DC. June 22–25.

Rader, K. 2004. *Making Mice: Standardizing Animals for American Biomedical Research, 1900–1955*. Princeton University Press. Princeton, NJ.

Ray, J. and S. Hansen. 2004. Temperament in the rat: sex differences and hormonal influences on harm avoidance and novelty seeking. *Behavioral Neuroscience* 118: 488–497.

Regan, T. 2001. The case for animals rights. In C. Cohen and T. Regan (eds.), *The Animal Rights Debate*. Rowman &Littlefield, Lanham, MD.

Reines, B. 1986. *Cancer Research on Animals: Impact and Alternatives*. National Anti-Vivisection Society, Chicago.

Robb, J. W. 1988. Can Animal Use be Ethically Justified? In H. N. Guttman, J. A. Mench, R. C. Simmonds (eds.), *Science and Animals: Addressing Contemporary Issues*. Conference of Scientists Center for Animal Welfare in Washington, DC. June 22–25.

Rodd, R. 1990. *Biology, Ethics, and Animals*. Clarendon Press. Oxford.

Rollin, B. 1981. *Animal Rights and Human Morality*. Prometheus Books, Buffalo.

Russell W. M. S. and R. L. Burch. 1959. *The Principles of Humane Experimental Technique*. Metheun, London. http://altweb.jhsph.edu/publications/humane_exp/het–toc.htm.

Sapolsky, R. M., and L. J. Share. 2004. A Pacific culture among wild baboons: Its emergence and transmission. Public Library of Science Biology 2(April):534–541. http://dx.doi.org/10.1371/journal.pbio.0020106.

Singer, P. 1975. *Animal Liberation*. Avon Books, New York.

Singer, P. 2004. Humans are sentient too. The Guardian, July 24, 2004. http://www.utilitarian.net/singer/by/20040730.htm

Sperling, S. 1988. *Animal Liberators: Research and Morality*. University of California Press. Berkeley.

U.S. Congress. Senate. Committee on Commerce. 1966. The Animal Welfare Act of August 24, 1966. 89th Congress S. Rep. 1281. http://www.nal.usda.gov/awic/legislat/pl89544.htm.

Wahrman, M. Z. 2002. *Brave New Judaism: When Science and Scripture Collide*. Brandeis University Press, Waltham, MA.

Waldau, P. 2002. *The Specter of Speciesism: Buddhist and Christian Views of Animals*. Oxford University Press, New York.

Wei, Q. and 9 others. 2004. Glucocorticoid receptor overexpression in forebrain: A mouse model of increased emotional lability. *Proceedings of the National Academy of Sciences USA* 101: 11851–11856.

Wescoat, J. L. Jr. 1998. The "Right of Thirst" for Animals in Islamic Law: A Comparative Approach. In J. Wolch and J. Emel (eds.), *Animal Geographies: Place, Politics, and Identity in the Nature-Culture Borderlands.* Verso Press, London.

White, L. 1967. The historical roots of our ecological crisis. *Science* 155: 1203–1207.

Williams, T. P. 1988. Ambient lighting and integrity of the retina. In H. N. Guttman, J. A. Mench, R. C. Simmonds (eds.), *Science and Animals: Addressing Contemporary Issues.* Conference of Scientists Center for Animal Welfare in Washington, DC. June 22–25.

Wolfe, C. 2003. *Animal Rites: American Culture, the Discourse of Species, and Posthumanist Theory.* University of Chicago Press. Chicago.

Wright, J. and C. Hensley. 2003. From animal cruelty to serial murder: Applying the graduation hypothesis. *International Journal of Offender Therapy and Comparative Criminology* 47: 71–88.

Zornberg, A. G. 1995. *The Beginning of Desire: Reflections on Genesis.* Doubleday, New York.

Zurlo, J., D. Rudacille and A. M. Goldberg. 2004. Animal and alternatives in testing: History, science, and ethics. Johns Hopkins Center for Alternatives to Animal Testing. http://caat.jhsph.edu/pubs/animal_alts/chap3.html

Index

About the Book

Editor: Andrew D. Sinauer

Project Editor: Carol J. Wigg

Photo Research: David McIntyre

Production Manager: Christopher Small

Book Design and Layout: Jefferson Johnson

Cover Design: Joanne Delphia and Jefferson Johnson

Index: Grant Hackett

Book and Cover Manufacture: Courier Companies, Inc.